BIOCHEMISTRY RESEARCH TRENDS

POLYUNSATURATED FATTY ACIDS (PUFAS)

FOOD SOURCES, HEALTH EFFECTS AND SIGNIFICANCE IN BIOCHEMISTRY

BIOCHEMISTRY RESEARCH TRENDS

Additional books in this series can be found on Nova's website under the Series tab.

Additional e-books in this series can be found on Nova's website under the e-Books tab.

BIOCHEMISTRY RESEARCH TRENDS

POLYUNSATURATED FATTY ACIDS (PUFAS)

FOOD SOURCES, HEALTH EFFECTS AND SIGNIFICANCE IN BIOCHEMISTRY

ANGEL CATALA
EDITOR

nova
science publishers
New York

Library of Congress Cataloging-in-Publication Data

ISBN: 978-1-53613-572-5

Published by Nova Science Publishers, Inc. † New York

CONTENTS

PREFACE

This book presents an overview of polyunsaturated fatty acids (pufas): food sources, health effects and significance in biochemistry. The topics analyzed cover a broad spectrum of polyunsaturated fatty acids and present new information in this area of research. The topics analyzed include: microalgae as a source of omega-3 polyunsaturated fatty acids; polyunsaturated fatty acids (pufas) in human health; polyunsaturated fatty acids in artificial membranes modeled by the langmuir monolayer technique; lipid peroxidation at the cross roads of cell proliferation and cancer; advancement in the knowledge of polyunsaturated fatty acids, from discovery to now; the discussion of whether or not polyunsaturated fatty acids are beneficial for failing hearts; polyunsaturated fatty acids (n3) of vegetal origin; proteomic approaches to understand the health effects of marine omega-3 pufa; essential fatty acids; and health implications and threats to sources.

Chapter 1 - The health benefits of omega-3 polyunsaturated fatty acids (n-3 PUFA) are general accepted and are mainly associated with the consumption of omega-3 long chain polyunsaturated fatty acids (n-3 LC-PUFA) containing a minimum of 20 carbon atoms, like eicosapentaenoic acid (EPA, C20:5 ω-3) and docosahexaenoic acid (DHA, C22:6 ω-3). Because of the well-known health benefits, there is an increased attention to the dietary intake of n-3 LC-PUFA. The traditional sources of n-3 LC-PUFA are fatty fish and fish oil, featured by a reducing stock.

Microalgae are unicellular aquatic organisms and could provide a sustainable alternative source for fatty fish and fish oil. There is a large diversity of microalgae, of which many are a primary producer of n-3 LC-PUFA. Depending on the species, the microalgae can be cultured photoautotrophically or heterotrophically. In general, they are characterized by an exponential growth and a high area productivity. Thereby, they show a limited competition with agriculture. Through these characteristics, microalgae are a promising source of n-3 LC-PUFA.

Before microalgae can be used into different food applications, harvesting and drying are crucial processing steps in the production of high quality n-3 LC-PUFA rich microalgal biomass. During the harvesting process, a large volume of water, originating from the growth medium, is removed to obtain a wet microalgal paste. This paste is susceptible to lipolytic and oxidative degradation. Therefore a supplemental drying process is being performed to obtain dry microalgal biomass.

Microalgal biomass can be incorporated into the food chain in different ways. Three different forms can be distinguished: direct incorporation of the whole biomass, direct incorporation of the oil extracted from the biomass and applications of the biomass and oil as an animal feed supplement. Past decade, research has been performed to create more insight in the enrichment of food products with n-3 LC-PUFA derived from microalgae. All photoautotrophic microalgae were incorporated as intact biomass, while heterotrophic microalgae were incorporated as oil. Moreover, the research on the incorporation of photoautotrophic and heterotrophic microalgae targets different food products. The focus for photoautotrophic biomass is mainly on biscuits and cereal products, in contrast to heterotrophic oil whereby egg, meat and fish products are the main products of interest. Both photoautotrophic and heterotrophic microalgae have shown their potential as an alternative source of n-3 LC-PUFA in different food products with each their advantages and disadvantages. Photoautotrophic biomass showed a higher oxidative stability than heterotrophic oil. On the other hand, the presence of pigments in photoautotrophic species resulted in larger changes in color of the enriched product compared to the heterotrophic species. However, addition of microalgae as a source of n-3 LC-PUFA to food products is an effective way to increase n-3 LC-PUFA levels in the diet.

Chapter 2 - Polyunsaturated fatty acids (PUFAs), which are known to be critical for human health, are lipids in which the constituent hydrocarbon chain possesses two or more carbon-carbon double bonds. PUFAs are distributed in almost all cells in the human body and affect membrane composition and function, eicosanoid synthesis, signal transduction, and gene expression. Linoleic acid (LA), an omega-6 fatty acid, and α-linolenic acid (ALA), an omega-3 fatty acid, are considered essential fatty acids because they cannot be synthesized by humans. The omega-3 fatty acids eicosapentaenoic acid (EPA) and docosahexaenoic acid (DHA) can be synthesized from ALA; however, due to low conversion efficiency, EPA and DHA should be obtained from food sources. Foods that provide omega-3 fatty acids include fish oil and certain plant and nut oils, whereas omega-6 fatty acids can be found in palm, soybean, rapeseed, and sunflower oils. Omega-9 fatty acids, which are not considered essential, can be found in animal fat and olive oil. Fish oil contains two omega-3 fatty acids, i.e., DHA and EPA. Some nuts, seeds, and vegetable oils contain ALA, which may be converted into DHA and EPA in the body. Importantly, omega-6 PUFAs have been shown to promote inflammation and augment many diseases, whereas omega-3 PUFAs appear to counter these adverse effects. Indeed, omega-3 fatty acids are thought to provide a wide range of health benefits, including reducing the risk of coronary heart disease and improving cholesterol levels. Studies of omega-3 fatty acids have shown promising effects in cancer, diabetes mellitus, depression, and attention-deficit hyperactivity disorder. Due to these potential health benefits, fish oil, which is rich in omega-3 fatty acids, has become a popular supplement. Moreover, recommended amounts of DHA and EPA, consumed from fish or fish oil supplements, may lower triglycerides and reduce the risk of heart attack, arrhythmia, and stroke in individuals with heart disorders.

To induce biological effects, PUFAs must be metabolized to their respective long-chain metabolites. PUFAs not only form precursors to respective prostaglandins (PGs), thromboxanes (TXs), and leukotrienes (LTs) but also modify lipoxins (LXs), resolvins, and isoprostanes. Certain PGs, TXs, and LTs have pro-inflammatory effects, whereas LXs and resolvins are anti-inflammatory molecules. Furthermore, PUFAs and their long-chain metabolites modulate the activities of angiotensin-converting enzymes (ACE) and HMG-CoA reductase enzymes, enhance acetylcholine levels in the brain, increase the synthesis of endothelial nitric oxide, augment diuresis, and enhance insulin action. Thus, PUFAs and their metabolites may function as endogenous ACE and HMG-CoA reductase inhibitors, nitric oxide enhancers, β-blockers, diuretics, antihypertensive, and anti-atherosclerotic molecules. Additionally, PUFAs and their long-chain metabolites affect signal transduction via activation of peroxisome proliferator-activated receptors. Thus, PUFAs and their derivatives are involved in many physiological and pathological actions to regain homeostasis.

Chapter 3 - The Langmuir monolayer technique is one of the methods used to model biomembranes and serves as a platform to study the interactions of membrane components with other biomolecules. This method, briefly reviewed in this paper, together with surface visualization method - Brewster angle microscopy (BAM) - has been applied to characterize six unsaturated fatty acids (UFAs), belonging to the omega-9 (oleic acid), omega-3 (α-linolenic and stearidonic acids) and omega-6 (linoleic, γ-linolenic and eicosadienoic acids) in Langmuir monolayers. The influence of such factors as subphase temperature, length of the aliphatic chain, degree of unsaturation and position of the double bonds on the conformation of molecules at the air/water interface and monolayer thickness has been discussed. This paper summarizes the results of interactions between unsaturated fatty acids and major membrane lipids, such as cholesterol and phospholipids. Strong interactions, leading to the formation of stable complexes of unsaturated fatty acids and cholesterol, in which free cholesterol is bound, have been found. In this way, UFAs may impact in lowering cholesterol level. Studies of interactions between unsaturated fatty acids and cholesterol/phospholipid mixed monolayers, mimicking cellular membranes, allowed to conclude that UFAs fluidize model membranes and - present in small amounts - increase their stability.

Chapter 4 - Polyunsaturated fatty acids (PUFAs) especially linoleic acid (LA, 18:2 n-6) and alpha-linolenic acid (ALA, 18:3 n-3) are widely distributed in various food items. Both LA and ALA are essential fatty acids (EFAs) that are present in various oils, leafy vegetables and other food items and are amenable to peroxidation during various processes including extraction, purification, and bleaching and cooking. Despite this, it is not clear what could be the biological consequences of lipid peroxides. In general, it is considered that lipid peroxides are harmful and are fraught with possible toxic effects on various tissues and cellular processes. Despite this belief, several studies revealed that lipid peroxides may have some beneficial actions on cell proliferation and cancer. It has been observed that formation of adequate amounts of lipid peroxides can inhibit cell proliferation especially those of cancer cells and smooth muscle cells of vascular tissue and thus may arrest cancer growth and prevent atherosclerosis. On the other hand, formation of excess of lipid peroxides, especially in cancer cells, could lead to their apoptosis and thus, regression of cancer. These and other results suggest that one of the important factors that has a regulatory role in cell proliferation and cancer is lipid peroxides. Based on these evidences, it is suggested that efforts made to selectively generate lipid peroxides in cancer cells may lead to their growth inhibition and elimination of cancer whereas if adequate amounts of peroxides form in smooth muscle cells of vascular tissue can arrest atherosclerosis. These studies suggest that a delicate balance need to be maintained in the metabolism of EFAs and their long-chain metabolites such as arachidonic (AA), eicosapentaenoic (EPA) and docosahexaenoic (DHA) acids and their conversion to lipid peroxides and formation of their respective pro-and anti-inflammatory products, which have a regulatory role in cancer and atherosclerosis and other diseases.

Chapter 5 - Polyunsaturated fatty acids (PUFAs) are fatty acids that contain more than one double bond in their backbone.

The most important polyunsaturated fatty acids can be grouped in two families: n-3 and n-6 fatty acids are a group of fatty acids, which have the first unsaturated bond in the third and sixth position respectively from the omega carbon.

The most important polyunsaturated fatty acids can be described as follow:

α-Linolenic acid C18:3n-3 is a polyunsaturated fatty acid (PUFA). It is a member of the group of essential fatty acids called n-3 fatty acids. α-Linolenic acid (18:3n-3) is the predominant plant-derived nutritional n-3 PUFA and is a precursor for docosahexaenoic acid 22:6n-3 (DHA).

Linoleic acid (18:2n-6) is the predominant plant-derived nutritional n-6 PUFA and is a precursor for arachidonic acid (20:4n-6) and eicosanoids.

The first evidences for the existence of polyunsaturated fatty acids containing a very high iodine number was first described in the late 1800s in studies of fish oils by and.

The first isolation and characterization of a tetraunsaturated 20-carbon fatty acid was reported by a young investigator at the Lister Institute, Percival Hartley, who was working on his D.Sc. degree from the University of London.

In 1929, George Oswald Burr and his wife Mildred, published a paper which discovered that elimination of fat from the diet of animals induced a deficiency illness, and their afterward papers showed that this illness could be prevented or cured by the addition of linoleic acid in the diet. Thus, they proved convincingly that linoleic acid was an essential fatty acid.

In this chapter the author would like to review some basic concepts of the chemistry and biochemistry of polyunsaturated fatty acids, and then the author will review some selective parts of the research they were involved with that range from the early sixties up to now.

Chapter 6 - Dietary consumption of polyunsaturated fatty acids (PUFAs), particularly when replacing saturated fatty acids, is a current behavioral modification strategy recommended by the American Heart Association/American College of Cardiology for the treatment and prevention of cardiovascular diseases. There is strong epidemiological data showing that both cardiovascular risk and mortality is reduced when

PUFAs are consumed. Although PUFAs have demonstrable benefits on serum lipid profiles, less is known about the direct effects on myocardial function, particularly in heart failure. In this review, a summary of the recent literature on the outcomes of PUFA consumption or supplementation on measures of cardiac function in both clinical and animal models of heart failure will be presented. As PUFAs are essential components of membranes, dietary modulation of membrane lipid composition, especially of the mitochondrial membrane, and its functional consequences will be explored. Finally, the role of PUFAs as signaling molecules to modulate both inflammatory and fibrosis pathways will be discussed.

Chapter 7 - It is well accepted that the consumption of very long-chain n-3 PUFA of marine origin have beneficial effects on plasma triglyceride concentrations and on systemic inflammation states. α-Linolenic acid (18: 3 n-3, ALA), although also a member of the n-3 PUFA family, is found predominantly in plants and its beneficial effects on health are more controversial. However, it has been suggested that ALA intake may reduce the risk of coronary heart disease when the intake of n-3 PUFA from marine sources is low. It is still unclear whether the effect is per se, or after its bioconversion to eicosapentaenoic (EPA) or docosahexaenoic (DHA) n-3 PUFA. Despite the recommendations of international institutions, including the World Health Organization, to increase fish consumption, in many countries it is difficult to achieve sufficient intake of n-3 PUFA. Access to n-3 PUFA of marine origin may be limited even in those regions with access to the sea. On the other hand, fish stocks are endangered around the world, making aquaculture the only sustainable solution for the future. Thus, the consumption of ALA-rich plant food is postulated as one of the most realistic alternatives to endure the n-3 PUFA intake. Among the different alternatives of plant n-3 PUFA-rich sources, chia (*Salvia hispanica*), sacha inchi (*Plukenetia volubilis*) and linseed (*Linum usitatissimum)*, have been considered. Oils obtained from their seeds can provide up to 65% of ALA. In addition, nuts are also appreciated sources of n-3 PUFA. Finally, transgenic plants have also been proposed as sustainable sources of EPA and DHA. The authors review here the current knowledge on plant n-3 PUFA consumption and the physiological mechanisms, mediated by triglyceride synthesis and the inhibition of inflammatory processes, by which they can exert beneficial effects on human health.

Chapter 8 - Marine ω-3 polyunsaturated fatty acids (PUFAs), i.e., eicosapentaenoic acid (EPA) and docosahexaenoic acid (DHA), have demonstrated their bioactivity in the improvement of human health, by preventing or even palliating numerous diseases, including diet-related diseases such as cardiovascular diseases (CVD) and type 2 diabetes, but also neurogenerative diseases. Proteomics, which is defined as the large-scale analysis of proteins, offers valuable information about structure, function, post-translational modifications, interactions and abundance of proteins, information that has contributed to explain mechanisms of this bioactivity of marine ω-3 PUFAs. The aim of this chapter is to present an overview of the diverse information available in the bibliography that the application of proteomics approaches has provided to understand critical issues regarding bioactivity and functional properties of marine EPA and DHA in human health. Therefore, this chapter summarizes the proteomics tools adopted to test the beneficial effects of EPA and DHA consumption, the disease studied and the main findings resulting from Proteomics. This information highlights the usefulness of Proteomics for studying the beneficial effects of marine ω-3 PUFAs consumption and can guide researchers to better design proteomics approaches to test the bioactivity of these marine PUFAs.

Chapter 9 - The importance of essential fatty acids (EFA) in supporting human health is well established. EFAs represent a structural and metabolic source of versatile agents necessary for membrane, cellular and trans-cellular function. While recommendations on n-3 and n-6 EFA in the diet have been defined by health regulatory bodies, population studies indicate that typical diets do not follow these recommendations. Insufficient dietary EFA contribute to many health problems such as cardiovascular disease and other morbidities. Current advances in biochemistry and biotechnology are improving knowledge of EFA activities and synthesis. Biochemical approaches may help to clarify the details of EFA interactions with metal ions, oxidative metabolites and transmembrane flux. Also, biotechnology production processes are improving the yield and sources of EFA, which, together with improved monitoring of fatty acids and their metabolites will allow improved understanding of dietary dynamics and effects on health. However, a serious problem is that the principal sources of EFA are threatened by environmental pollution. Without the intervention of current food production, processing and legislation to limit environmental pollution, it is likely that insufficient EFA supplies will be available to support a healthy global population.

In: Polyunsaturated Fatty Acids (PUFAs)
Editor: Angel Catala
ISBN: 978-1-53613-572-5

Chapter 1

MICROALGAE AS A SOURCE OF OMEGA-3 POLYUNSATURATED FATTY ACIDS

Lore Gheysen[1,2,*], Valerie Matton[1,2,†] and Imogen Foubert[1,2,‡]
[1]KU Leuven Kulak, Research Unit Food and Lipids, Kortrijk, Belgium
[2]Leuven Food Science and Nutrition Research Centre (LFoRCe), KU Leuven, Leuven, Belgium

ABSTRACT

The health benefits of omega-3 polyunsaturated fatty acids (n-3 PUFA) are general accepted and are mainly associated with the consumption of omega-3 long chain polyunsaturated fatty acids (n-3 LC-PUFA) containing a minimum of 20 carbon atoms, like eicosapentaenoic acid (EPA, C20:5 ω-3) and docosahexaenoic acid

* Email: lore.gheysen@kuleuven.be.
† Email: valerie.matton@kuleuven.be.
‡ Corresponding Author Email: imogen.foubert@kuleuven.be.

(DHA, C22:6 ω-3). Because of the well-known health benefits, there is an increased attention to the dietary intake of n-3 LC-PUFA. The traditional sources of n-3 LC-PUFA are fatty fish and fish oil, featured by a reducing stock.

Microalgae are unicellular aquatic organisms and could provide a sustainable alternative source for fatty fish and fish oil. There is a large diversity of microalgae, of which many are a primary producer of n-3 LC-PUFA. Depending on the species, the microalgae can be cultured photoautotrophically or heterotrophically. In general, they are characterized by an exponential growth and a high area productivity. Thereby, they show a limited competition with agriculture. Through these characteristics, microalgae are a promising source of n-3 LC-PUFA.

Before microalgae can be used into different food applications, harvesting and drying are crucial processing steps in the production of high quality n-3 LC-PUFA rich microalgal biomass. During the harvesting process, a large volume of water, originating from the growth medium, is removed to obtain a wet microalgal paste. This paste is susceptible to lipolytic and oxidative degradation. Therefore a supplemental drying process is being performed to obtain dry microalgal biomass.

Microalgal biomass can be incorporated into the food chain in different ways. Three different forms can be distinguished: direct incorporation of the whole biomass, direct incorporation of the oil extracted from the biomass and applications of the biomass and oil as an animal feed supplement. Past decade, research has been performed to create more insight in the enrichment of food products with n-3 LC-PUFA derived from microalgae. All photoautotrophic microalgae were incorporated as intact biomass, while heterotrophic microalgae were incorporated as oil. Moreover, the research on the incorporation of photoautotrophic and heterotrophic microalgae targets different food products. The focus for photoautotrophic biomass is mainly on biscuits and cereal products, in contrast to heterotrophic oil whereby egg, meat and fish products are the main products of interest. Both photoautotrophic and heterotrophic microalgae have shown their potential as an alternative source of n-3 LC-PUFA in different food products with each their advantages and disadvantages. Photoautotrophic biomass showed a higher oxidative stability than heterotrophic oil. On the other hand, the presence of pigments in photoautotrophic species resulted in larger changes in color of the enriched product compared to the heterotrophic species. However, addition of microalgae as a source of n-3 LC-PUFA to food products is an effective way to increase n-3 LC-PUFA levels in the diet.

1. Health Benefits of n-3 LC-PUFA

The health benefits of omega-3 polyunsaturated fatty acids (n-3 PUFA) are generally accepted and are mainly associated with the consumption of omega-3 long chain polyunsaturated fatty acids (n-3 LC-PUFA) containing a minimum of 20 carbon atoms, like eicosapentaenoic acid (EPA, C20:5 n-3) and docosahexaenoic acid (DHA, C22:6 n-3) (Calder, 2014). EPA and DHA respectively contain five and six double bounds, of which the first one is located on the third C-atom counting from the methyl end. Both fatty acids can be formed from the essential fatty acid alpha-linolenic acid (ALA, C18:3 n-3), which must be acquired through the diet, making EPA and DHA semi-essential fatty acids. A chain of desaturase and elongase enzymes in the human body is able to transform ALA into EPA and DHA (Russo, 2009). This conversion is however very limited with conversion factors stated in literature varying between 0.01 and 10% (Swanson et al., 2012). The reason is that the pathway for the biosynthesis of n-3 LC-PUFA experiences competition from the linoleic acid (LA, C18:3 ω-6) conversion to long chain omega-6 fatty acids (n-6 LC-PUFA) as both pathways are using the same enzymes for the biosynthesis of their corresponding long chain fatty acids (Ryckebosch et al., 2012). The last decades the ratio n-6:n-3 in the western diet has become unbalanced with ratios typically 10:1 to 25:1 instead of the recommended 2:1 to 6:1 (Brasseur et al., 2004; Simopoulos, 2002; Zárate et al., 2017). The direct consumption of EPA and DHA is indispensable, considering their low *in vivo* synthesis from ALA (Molendi-Coste et al., 2011).

After consumption, EPA and DHA are mainly incorporated in the phospholipids of the cell membranes, where they can influence the membrane fluidity. This leads to a change in membrane protein activity, an impact on the hormone-receptor connection and the transport and signal regulation (Calder, 2014; Russo, 2009; Ryckebosch et al., 2012). EPA and DHA are involved in the formation of precursors of anti-inflammatory immune responses and bioactive eicosanoids (Calder, 2014), while n-6 LC-PUFA lead to the formation of pro-inflammatory eicosanoids. A higher intake of n-3 LC-PUFA helps to restore the n-6:n-3 balance and therewith the suppression of pro-inflammatory compounds (Covington, 2004). The European Food Safety Authority (EFSA) approved different health claims related to the consumption of EPA and DHA. A combined consumption of EPA and DHA contributes to normal functioning of the heart and cardiac related functions. Thereby, a normal blood pressure and a normal blood concentration of triglycerides is maintained. Consumption of only DHA contributes to the maintenance of normal vision and the development of a normal brain function. Maternal intake of DHA helps to the development of the eye of the fetus and breastfed infants (EFSA, 2011). In literature, health benefits of EPA and DHA related to depression, dementia, rheumatoid arthritis, kidney functioning…, are described as well (Swanson et al., 2012). However these benefits are not yet approved by EFSA. Worldwide, several organizations (The World Health Organization (WHO) and governments around the world) have set up a recommended daily intake of 250-500 mg of EPA and DHA based on the results on the prevention of cardiovascular disorders and cognitive development. Higher intakes are advised for pregnant and lactating woman (EFSA, 2011; Flock et al., 2013; GOED, 2016; Kris-Etherton et al., 2009). Long-term intake of combined doses of EPA and DHA up to 5 g/day, and supplemental intakes of EPA alone up to 1.8 g/day, do not raise safety concerns for adults (EFSA, 2011).

Food naturally rich in n-3 LC-PUFA, n-3 LC-PUFA enriched food and n-3 LC-PUFA supplements can attribute to consuming a higher level of n-3 LC-PUFA. Supplements only reach a small part of the population (Sioen et al., 2010). The consumption of foods naturally rich in n-3 LC-PUFA, fatty fish and fish oil, is low in most countries and a change in diet would be needed to meet the recommended intake (Molendi-Coste et al., 2011; Sioen et al., 2010). In contrast, no change in eating habits is needed to enhance the intake of n-3 LC-PUFA via enriched food products, which reach a larger part of the population (Anbudhasan et al., 2014). In this respect, there is an increased interest to enrich food products with n-3 LC-PUFA.

Especially fatty fish, like herring and salmon are sources rich in EPA and DHA, with contents up to 1700 mg/100 g (Kris-Etherton et al., 2000;

Russo, 2009). Also other marine sources, like krill and squid contain EPA and DHA (Deckelbaum & Torrejon, 2012). However, fish, krill and squid do not produce n-3 LC-PUFA themselves, since they are lacking enzymes for the conversion of ALA to EPA and DHA. Besides, there is accumulation of PCB's, dioxins and lead in the marine food chain as well (Sidhu, 2003; Walker et al., 2015). Moreover, the current fish stock is too low to provide the global population with the recommended daily intake. An alternative and sustainable source of n-3 LC-PUFA is thus needed. Microalgae are the primary producers of n-3 LC-PUFA, which accumulate along the food chain and can provide such an alternative (Matos et al., 2017; Ryckebosch et al., 2012).

2. Microalgae as an Alternative Source

Microalgae are unicellular aquatic organisms, growing in marine or fresh water. Most of them are eukaryotic and they exist individually, in chains or in groups (Buono et al., 2014). Microalgae are featured by a large diversity, with an estimation of more than 200.000 species (Jasvinder & Saxena, 2015; Salim et al., 2014) differing amongst others in shape, size and cell wall composition (Hillebrand & Sommer, 1999).

Depending on the species, microalgae can be cultured photoautotrophically, heterotrophically or mixotrophically. The photoautotrophic growth of microalgae requires only light and CO_2 as an inorganic carbon source to manufacture organic material. Furthermore, some nutrients like nitrogen, sulphur and phosphorus are stimulating the development. Via their photosynthetic pathways, the captured light and CO_2 is converted into energy (Buono et al., 2014; Pignolet et al., 2013; Ryckebosch et al., 2012). Heterotrophic microalgae grow by fermentation on organic substrates without the presence of light (Morales-Sánchez et al., 2014). Some species can be grown mixotrophically, which refers to their ability to perform photosynthesis and fermentation on organic carbon sources simultaneously.

Eukaryotic photoautotrophic microalgae are recognized as primitive plants, as their general cell structure is comparable with plant cells. A rigid cell wall surrounds the cell and forms the outside of the algal cell. The membrane located on the inner side of the cell wall functions as a selective barrier. According to plant cells, the cytosol includes different cell organelles like mitochondria, nucleus, vacuole, chloroplast, Golgi apparatus, and endoplasmic reticulum. Also lipid bodies, which store the generated energy, are present in the cytosol (Jasvinder & Saxena, 2015; Pignolet et al., 2013). Eukaryotic heterotrophic microalgae, are thin-walled vegetative cells containing mitochondria, nucleus, vacuole, Golgi apparatus, endoplasmic reticulum and lipid bodies (Yokoyama & Honda, 2007).

The chemical composition of microalgal cells shows a big variability depending on the growth conditions such as water temperature, nutrient availability, salinity and light intensity (Buono et al., 2014). Furthermore, the chemical composition differs depending on the species. Microalgae can be found with a wide range of protein (20-71%), carbohydrate (8-64%) and lipid (2-70%) concentrations (Chacón-Lee & González-Mariño, 2010). Microalgae growing under normal conditions produce significant amounts of lipids, neutral as well as polar lipids (phospholipids and glycolipids), in the stationary growth phase (Gnouma et al., 2017). Most of the polar lipids are located in the cell membrane and the cell organelle membranes, while the neutral lipids function as storage lipids (Hu et al., 2008; Pignolet et al., 2013). Moreover, microalgae often contain high-value products, like n-3 LC-PUFA and antioxidants, which can be used as functional ingredient. Many microalgal species are primary producers of n-3 LC-PUFA, such as EPA and DHA, in the marine food chain (Adarme-Vega et al., 2012). They regulate the fluidity and membrane function in the microalgal cell (Bellou et al., 2016; Matos et al., 2017; Salim et al., 2014).

Table 1 reports the EPA and DHA concentrations (as % of total fatty acids) of different microalgal species (based on Ryckebosch et al., 2012).

Table 1. Overview of total lipids (%), EPA and DHA concentrations (as% of total acids) in different microalgal species (based on Ryckebosch et al., 2012)

Phylum	Species	Total lipids	EPA	DHA
Bacillariophyta	*Chaetoceros*	9-40	8-22	0.03-5
	Phaeodactylum	15-40	2-36	0-3
	Skeletonema	3-25	13-36	1-6
	Thalassiosira	9-26	11-17	1-5
Chlorophyta	*Tetraselmis*	6-26	4-11	0-0.1
Cryptophyta	*Cryptomonas*	12-22	3-25	2.5-10
	Rhodomonas	7-19	8-18	4-9
Dinoflagellata	*Crypthecodinium*	30-50	0-1	47-97
Haptophyta	*Diacronema*	7-36	5-29	4-19
	Isochrysis	7-33	1-27	1-40
Heterokontophyta	*Nannochloropsis*	16-68	9-38	0.03-3
	Schizochytrium	50-77	0-2	20-53
Rhodophyta	*Porphyridium*	11-18	2-38	0-0.2

These high amounts of n-3 LC-PUFA make microalgae a reliable alternative source for fish and fish oil. Furthermore, microalgae have different advantages compared to fish as the traditional n-3 LC-PUFA source. Microalgal organisms grow exponentially and this high growth rate allows to collect large amounts of biomass on a limited time. They are cultivated in reactors or open pond systems, of which the conditions can be controlled, leading to a more constant biochemical composition and the absence of chemical contaminants. Moreover, no fertile land is needed and microalgae fit in a vegetarian, and even vegan, diet (Ryckebosch et al., 2012). Besides, photoautotrophic microalgae contain a significant amount of their EPA and DHA in the polar lipid classes, which may lead to a higher bioavailability compared to n-3 LC-PUFA in the neutral lipids. Moreover, a higher oxidative stability was observed in photoautotrophic microalgae, which is probably explained by the presence of endogenous antioxidants (Matos et al., 2017; Ryckebosch et al., 2013).

3. Production Chain of Microalgae

Before high quality n-3 LC-PUFA rich microalgal biomass can be used in food applications, different steps in the production process should be completed. In general the production process is executed as shown in Figure 1 (Vandamme, 2016).

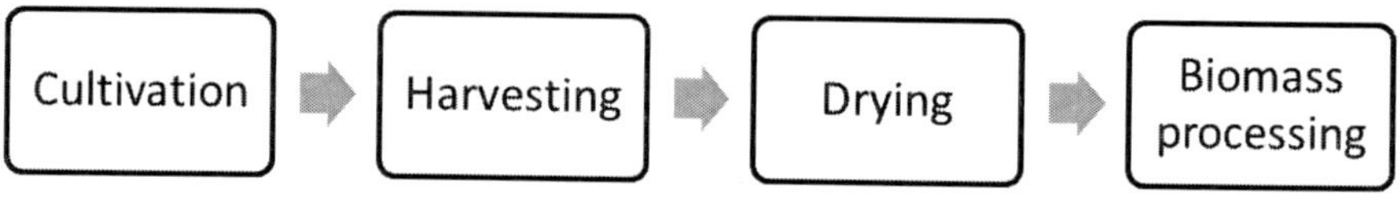

Figure 1. Production process of microalgae and their derived products (based on Vandamme, 2016).

Photoautotrophic microalgae are mainly cultivated in open raceway ponds or in closed photobioreactors. Open raceway ponds are characterized by a lower operational cost and an easier handling while upscaling the production. However, they are less controlled and therewith less resistant to contamination compared to closed photobioreactors. Usually, a cell density of 0.5 g/L in open pond reactors and up to 5 g/L in closed photobioreactors can be obtained (Dineshkumar et al., 2017; Suali & Sarbatly, 2012). Heterotrophic microalgae are typically grown in a conventional fermentor, whereby a cell density of 25-150 g/L can be achieved (Bumbak et al., 2011; Chen et al., 2011). The cultivation curve of microalgae proceeds according to a fixed pattern. First, microalgae need some time to adapt to their new conditions. In this phase, called the lag phase, there is almost no growth. This period of adaptation is followed by the exponential phase, where microalgae multiply rapidly and the cell density increases (Chen & Jiang, 2017; Gnouma et al., 2017). In the stationary phase, nutrients become limited causing a smaller cell density increase (Pires, 2015). For most microalgal species, the highest concentrations of EPA and DHA (as % of total fatty acids) are obtained in the stationary phase (Boelen et al., 2017).

During the harvesting process, a large volume of water, originating from the growth medium, is removed to obtain a wet microalgal paste (Borowitzka & Moheimani, 2013). Different harvesting techniques can be used, amongst others centrifugation, filtration, flotation, flocculation and sedimentation (Suali & Sarbatly, 2012). Centrifugation is the common harvesting method since it is rapid, efficient and suitable for most species, but on the other hand, it is an expensive and energy-intensive way to separate the microalgae from their growth medium. Therefore in the past years, more research on the alternative and low cost harvesting methods is being performed (Milledge & Heaven, 2013; Vandamme et al., 2013).

The paste obtained by harvesting (15-25% dry weight) is susceptible to lipolytic degradation by endogenous hydrolytic enzymes as Balduyck et al. (2016) observed that free fatty acids are formed very rapidly during post-harvest storage of wet biomass of *Isochrysis* at 20°C and 4°C. The lipolysis rate is however strongly influenced by the cell integrity, depending on species and pretreatment (Balduyck et al., 2017). Free fatty acids oxidize more easily compared to esterified fatty acids (Frankel, 1980) and therefore a supplemental drying process is performed to obtain stable and dry microalgal biomass. Commonly, freeze drying or spray drying are the techniques used (Ryckebosch et al., 2011).

Once dried, the microalgal biomass can be incorporated into the food chain in different ways. Three different delivery systems can be distinguished: direct incorporation of the whole biomass, direct incorporation of oil extracted from the biomass and indirect application as an animal feed supplement (Buono et al., 2014). When incorporating the biomass, no further downstream processing is needed. For the incorporation of n-3 LC-PUFA oil, a supplemental extraction step should be executed (Mercer & Armenta, 2011). Both, microalgal biomass and oil, can be used as an animal feed supplement as well. This may lead to an indirect increase of n-3 LC-PUFA in the food chain via animal derived products like eggs or meat (Fraeye et al., 2012; Madeira et al., 2017). Indirect incorporation will not be discussed further in this chapter.

4. Direct Incorporation of n-3 LC-PUFA Rich Microalgae into the Food Chain

During the past decade, there has been an increased interest to enrich food products with n-3 LC-PUFA, since no change in eating habits is then needed to enhance their intake and as such a large part of the population is attained (Anbudhasan et al., 2014). Research has been performed to create more insight in the enrichment of food products with n-3 LC-PUFA derived from microalgae. Table 2 presents an overview of the literature on the direct incorporation of n-3 LC-PUFA derived from microalgal species in food products. This overview summarizes the studies, chronologically per microalgal species, focusing on microalgal species cultivated by photoautotrophic and heterotrophic growth. A clear difference in the used delivery system of photoautotrophic and heterotrophic microalgae can be observed. All photoautotrophic microalgae were incorporated as intact biomass, while heterotrophic microalgae were incorporated as oil. Only biomass of *Tetraselmis* and oil extracted from *Schizochytrium* and *Crypthecodinium* is labeled as General Recognized As Save (GRAS) by the Food and Drug Administration (FDA) and food grade by EFSA. This explains the focus on the incorporation of photoautotrophic biomass on the one hand and heterotrophic microalgal oil on the other hand (EFSA, 2011; FDA, 2014). The research on the incorporation of photoautotrophic and heterotrophic microalgae targets different food products. The focus for photoautotrophic biomass is mainly on biscuits and other cereal products, in contrast to heterotrophic oil whereby egg, meat and fish products are the main products of interest. The amount of added biomass or oil and the final concentration of n-3 LC-PUFA in the food product is presented (Table 2) as well. These numbers were provided by the authors or calculated, if possible. A large variation of n-3 LC-PUFA addition can be observed. The claim of a food product 'rich in omega-3' demands at least 80 mg of EPA + DHA per 100 g or per 100 kcal, whichever is limiting (European Commission, 2010). The incorporation of the heterotrophic oil is often overdosed compared to the claim. For each of the studies the main focus of the experiment is indicated as well. For photoautotrophic microalgae, it can be observed that besides the amount of n-3 LC-PUFA, mainly sensorial characteristics, texture and color were evaluated. The number of studies focusing on the impact of food processing steps and lipid oxidation are very limited. The opposite could be observed for studies using heterotrophic oil, where a strong focus on lipid oxidation is observed. Sensorial characteristics, texture and color were also often taken into account. Furthermore, it was observed that the studies dealing with the incorporation of photoautotrophic biomass on the one hand and heterotrophic oil on the other, are executed by a limited number of research groups. The results of the different researches will be discussed per microalgal species in detail below.

4.1. Photoautotrophic Microalgae

4.1.1. *Diacronema* sp.

The brownish-green microalga *Diacronema* sp., formerly called *Pavlova* and taxonomically belonging to *Haptophyceae*, is a marine microalga containing EPA and DHA (Bendif et al., 2011). The potential of this species was screened in gelled desserts as a dairy analog and pasta respectively by Gouveia et al. (2008a) and Fradique et al. (2013).

4.1.1.1. n-3 LC-PUFA and Processing

The authors of both studies incorporated similar amounts of *Diacronema* biomass in varying concentrations between 0.1 and 2%. As expected, this led to an increase in n-3 LC-PUFA compared to the control in all samples (Gouveia et al., 2008a; Fradique et al., 2013). Fradique et al. (2013) reported 20, 40 and 80 mg n-3 LC-PUFA/100 g pasta by the addition of respectively 0.5, 1 and 2% microalgal biomass. These amounts are in line with regulations to claim the product as a source of n-3 LC-PUFA (40 mg) or rich in n-3 LC-PUFA (80 mg). Gouveia et al. (2008a) only presented relative amounts of EPA and DHA, which were not sufficient to calculate absolute amounts of n-3 LC-PUFA in the final product.

Table 2. Overview of research on the direct incorporation of n-3 LC-PUFA derived from microalgae in food products

Growth	Delivery system	Microalgal species	Food product	Amount of addition (%)	Amount of n-3 LC-PUFA (mg/100 g product)	Focus on						References
						n-3 LC-PUFA	Processing	Lipid Oxidation	Sensorial characte-ristics	Texture	Color	
Photo-autotrophic	Biomass	*Diacronema* sp.	Dairy analogs	0.1/0.25/ 0.5/0.75		x	x			x	x	Gouveia et al., 2008a
			Pasta	0.5/1/2	20/40/80	x	x		x	x	x	Fradique et al., 2013
		Isochrysis sp.	Biscuits	1/3	101/310	x	x			x	x	Gouveia et al., 2008b
			Pasta	0.5/1/2	30/60/120	x			x			Fradique et al., 2013
			Chikkis	1/3	102/320	x		x	x	x	x	Babuskin et al., 2015
			Bread	0.4						x	x	Garcia-Segovia et al., 2017
		Nanno-chloropsis sp.	Pasta Biscuits	0.5/1/3 1/2/3	31/63/190 98/197/298	x			x	x	x	Babuskin et al., 2014
			Chikkis	1/3	75/240	x		x	x	x	x	Babuskin et al., 2015
			Bread	0.4						x	x	Garcia-Segovia et al., 2017

Growth	Delivery system	Microalgal species	Food product	Amount of addition (%)	Amount of n-3 LC-PUFA (mg/100 g product)	Focus on						References
						n-3 LC-PUFA	Processing	Lipid Oxidation	Sensorial characte-ristics	Texture	Color	
		Phaeo-dactylum sp.	Biscuits	2/6						x	x	Batista et al., 2017
		Tetraselmis sp.	Bread	0.4						x	x	Garcia-Segovia et al., 2017
Hetero-trophic	Oil	*Crypthe-codinium* sp.	Surimi		588	x		x		x	x	Park et al., 2004
			Yoghurt	0.18	147			x	x			Chee et al., 2005
			Milk	2/5/10	800/2000/4000			x	x			Gallaher et al., 2005
			Patties, sausages, ham		454	x	x	x	x		x	Lee et al., 2006
			Sausages	1.1	400	x						Lopez-Lopez et al., 2009
			Sausages	10	3500	x		x	x		x	García-Íñiguez de Ciriano et al., 2010
			Egg sticks	10	4000	x						Kassis et al., 2010
			Surimi		588	x		x		x	x	Pietrowski et al., 2011
			Egg sticks	10	4000			x	x	x	x	Sedoski et al., 2012

Table 2. (Continued)

Growth	Delivery system	Microalgal species	Food product	Amount of addition (%)	Amount of n-3 LC-PUFA (mg/100 g product)	Focus on						References
						n-3 LC-PUFA	Processing	Lipid Oxidation	Sensorial characte-ristics	Texture	Color	
			Patties	0.1	24.4	x	x	x	x			Alejandre et al., 2017
		Schizo-chytrium sp.	Bread	0.325/0.65	78/156	x			x	x		Serna-Saldivar et al., 2006
			Sausages	15/25	1340/1490	*x*		x	x			Valencia et al., 2007
			Surimi	0.5/1	200/400	x		x		x		Tolasa et al., 2010
			Sausages	1.32	450	x		x	x	x	x	Berasategi et al., 2014
			Egg sticks	0.35	140	x		x	x		x	Lamas et al., 2015

No significant impact of cooking the pasta (3-5 min in boiling water) or heating the dessert to gelling temperatures (5 min at 90°C) on the n-3 LC-PUFA concentration was reported by the authors of both studies. This indicates a thermal resistance of EPA and DHA during processing of these food products (Gouveia et al., 2008; Fradique et al., 2013).

4.1.1.2. Lipid Oxidation

No lipid oxidation parameters were followed in the studies.

4.1.1.3. Sensorial Characteristics, Texture and Color

At levels of 80 mg n-3 LC-PUFA addition (2% biomass) per 100 g pasta, an untrained panel perceived a slight depreciative fish flavor. They marked the strange flavor with 7 on a scale of 9 compared to a mark of 2 for the control and incorporation concentrations of 20 mg n-3 LC-PUFA/100 g (0.5% biomass). Furthermore, on the global appreciation the panel showed a preference for the control pasta (almost 7 on a scale of 10), although the enriched pasta (0.5 and 1%) was positively classified (5.5 on a scale of 10). The highest incorporation concentration (2%) was marked with a score of 3 and was thus unacceptable (Fradique et al., 2013).

The texture of the pasta was not influenced by the incorporation of the microalgal biomass. The same mark of almost 7 (on a scale of 10) was given for the enriched pasta as well as for the control (Fradique et al., 2013). On the contrary, the gelled dessert reinforced by addition of higher incorporation concentrations, which is undesirable for such food products (Gouveia et al., 2008a).

The score for the color did not depend on incorporation concentration, but was always lower compared to the control (Gouveia et al., 2008a; Fradique et al., 2013).

4.1.2. *Isochrysis* sp.

Isochrysis sp., with some strains currently named as T-*Isochrysis*, is a brownish-golden marine microalga containing EPA and DHA (Bendif et al., 2013; Ryckebosch et al., 2014). Normally the presence of EPA in the fatty acid profile is rather limited and thus larger amounts of EPA probably indicate contamination during cultivation (Hubert et al., 2017; Nalder et al., 2015). This species is linked to the phylum *Haptophyta.* The potential of this species was screened in biscuits, pasta, chikkis (a traditional Indian sweet generally made from groundnuts and jaggery) and bread in studies of Gouveia et al. (2008b), Fradique et al. (2013), Babuskin et al. (2015) and Garcia-Segovia et al. (2017).

4.1.2.1. n-3 LC-PUFA and Processing

The microalgal biomass of *Isochrysis* was incorporated in the food products in concentrations between 0.4 and 3% (Gouveia et al., 2008b; Fradique et al., 2013; Babuskin et al., 2015; Garcia-Segovia et al., 2017) logically leading to an increased n-3 LC-PUFA content compared to the control varying between 30 and 310 mg/100 g product (Gouveia et al., 2008b; Fradique et al., 2013; Babuskin et al., 2015). In the studies of Gouveia et al. (2008b), Fradique et al. (2013) and Babuskin et al. (2015) resulted the addition of 1% biomass in different amounts of n-3 LC-PUFA in the final product, respectively 101 mg n-3 LC-PUFA/100 g biscuits, 60 mg n-3 LC-PUFA/100 g pasta and 102 mg n-3 LC-PUFA/100 g chikkis. The observed differences may be explained by the different ways to prepare the food products and batch variation of the microalgal species, and thus the composition of the used microalgae. Garcia-Segovia et al. (2017) presented no results about the fatty acid profile of the microalgae nor the amount of n-3 LC-PUFA in the final food product.

Gouveia et al. (2008b) only determined the n-3 LC-PUFA content after the baking process (30 min at 180°C). As expected, the amount of n-3 LC-PUFA was 3 times higher in the biscuits with 3% incorporated compared to 1%. Based on these results the authors suggested that the encapsulation of the n-3 LC-PUFA by the cells could protect them against thermal degradation. However, the amount of n-3 LC-PUFA of the biscuits before the baking process was not determined.

4.1.2.2. Lipid Oxidation

Babuskin et al. (2015) stored the chikkis enriched with 3% biomass for 60 days at 27°C. Natural (rosemary extract) or synthetic (BHA) antioxidants were added in different concentrations (1-5%). The peroxide value of the control (3% biomass without supplemental antioxidants) increased significantly after 30 days of storage, while the value of the chikkis with additional antioxidants did not. The peroxide value was determined by an iodometric titration, which is not the optimal method in colored samples like photoautotrophic microalgae enriched food products. Besides, it is suggested to follow both primary as well as secondary oxidation. The authors concluded that these results demonstrate the need of additional antioxidants in the food product, although the increase in the peroxide value of the control (between zero and 60 days) is still reasonably low (20 meq active oxygen/kg) and comparable with the results of Ryckebosch et al. (2013), who stored *Isochrysis* oil for 8 weeks at 37°C. Ryckebosch et al. (2013) observed, in contrast, an increase in the peroxide values for commercial n-3 LC-PUFA sources up to 300 meq active oxygen/kg (between zero and 14 days).

4.1.2.3. Sensorial Characteristics, Texture and Color

Similar results for the sensorial characteristics were observed for pasta and chikkis enriched with n-3 LC-PUFA derived from *Isochrysis* in the study of Fradique et al. (2013) and Babuskin et al. (2015) respectively. Food products without microalgae were preferred compared to those enriched with microalgae, although the panelists have classified the enriched products as 'acceptable'. Moreover, the incorporation concentration had a significant impact on the global appreciation. Pasta and chikkis enriched with 1% microalgal biomass were marked with 6 and 7.1, while the control obtained respectively 7 and 8.2. Products enriched with a higher concentration (up to 3%) showed a lower global appreciation of 4 and 4.9, because a fishy flavor was observed at these higher incorporation concentrations.

The texture of pasta, chikkis and bread was not modified by the incorporation of *Isochrysis* biomass compared to the control, while the firmness of the biscuits increased linearly with the microalgal biomass concentration (Gouveia et al., 2008b; Fradique et al., 2013; Babuskin et al., 2015; Garcia-Segovia et al., 2017).

The color of the enriched products was scored lower compared to the control and was limitedly influenced by the incorporation concentration. Decreased L*, a* and b* values were observed, which can be ascribed to the pigments of the microalgal biomass (Gouveia et al., 2008b; Fradique et al., 2013; Babuskin et al., 2015; Garcia-Segovia et al., 2017). In the studies of Gouveia et al. (2008b) and Babuskin et al. (2015) enriched food products were stored for respectively 12 and 8 weeks and within a sample no color changes could be detected over time.

4.1.3. *Nannochloropsis* sp.

The greenish, marine microalga *Nannochloropsis* sp., belonging to the *Haptophyceae* and also named as *Microchloropsis*, contains high amounts of EPA (Fawley et al., 2015; Rocha et al., 2003). The potential of *Nannochloropsis* sp. was screened in chikkis, pasta, biscuits and bread by Babuskin et al. (2014), Babuskin et al. (2015) and Garcia-Segovia et al. (2017).

4.1.3.1. n-3 LC-PUFA and Processing

Food products were enriched with 0.4 to 3% of *Nannochloropsis* biomass (Babuskin et al., 2014 and 2015; Garcia-Segovia et al., 2017). Babuskin et al. (2014 and 2015) observed an n-3 LC-PUFA concentration of 63, 75 and 198 mg/100 g product for respectively pasta, chikkis and biscuits enriched with 1% biomass and 190, 240 and 298 mg/100 g product for an incorporation concentration of 3%, while no n-3 LC-PUFA was present in the control. Obviously, higher incorporation concentrations resulted in linearly higher levels of n-3 LC-PUFA in the final product. Pasta, chikkis and biscuits were prepared in a different way, which can explain the differences of the n-3 LC-PUFA concentration in the final product when adding the same amount of biomass (1 or 3%). No results about the fatty acid profile of the enriched bread were given (Garcia-Segovia et al., 2017).

None of the studies focused on the impact of processing on the n-3 LC-PUFA concentration in the product.

4.1.3.2. Lipid Oxidation

A 60 days storage experiment executed by Babuskin et al. (2015) at 27°C with 3% *Nannochloropsis* enriched chikkis with and without added (natural or synthetic) antioxidants resulted in an increase of the peroxide value after 15 days for all samples. The control (without added antioxidants) showed the highest increase in the peroxide value (20 meq active oxygen/kg) between zero and 60 days, while for the chikkis with antioxidants an increase between 3 and 10 meq active oxygen/kg was observed. As already explained above for Isochrysis, the authors stated, based on these results, that extra antioxidants were needed for the prevention of lipid oxidation in enriched chikkis. However, the increase in the peroxide value is still low compared with the results of the commercial n-3 LC-PUFA sources of Ryckebosch et al. (2013).

4.1.3.3. Sensorial Characteristics, Texture and Color

The sensorial characteristics of enriched chikkis, pasta and cookies showed similar results in the studies of Babuskin et al. (2014 and 2015). The overall acceptability of food products without incorporation was preferred with a mark of respectively 7.9, 7.9 and 7 on a scale of 9. However, incorporation concentrations up to 2% for cookies and up to 3% for chikkis and pasta were still positively classified (i.e., with a mark higher than 4.5), while a fishy taste was observed by the panelist in cookies with an incorporation concentration of 3% probably explaining the lower global appreciation (3.3) of these products.

The firmness of the cookies and pasta increased with a higher incorporation concentration, while the texture of chikkis and bread was not affected by the used incorporation concentrations (Babuskin et al., 2014 and 2015; Garcia-Segovia et al., 2017). Textural changes during storage within a sample were comparable for all enriched samples as well as for the control (Babuskin et al., 2014 and 2015).

The color of the food products (chikkis, pasta, biscuits and bread) changed by incorporation of *Nannochloropsis*. The lightness value L* for these products decreased with a factor 1.2 to 1.7 compared to the control, meaning the enriched products were darker. In none of the products, the L* value was influenced by the incorporation concentration. The a* value decreased dramatically (less than half of the value of the control) in all enriched food products and a larger decrease was observed at higher incorporation concentrations. The b* value was only limitedly influenced by the addition of *Nannochloropsis.* In pasta a slight increase of the b* value could be observed, while in bread no significant difference and in chikkis and cookies a slight decrease was detected (Babuskin et al., 2014 and 2015; Garcia-Segovia et al., 2017). Furthermore, no significant impact of the incorporation concentration could be detected. In the studies of Babuskin et al. (2014 and 2015) storage of the incorporated food products for 8 weeks showed no color changes over time.

4.1.4. *Phaeodactylum* sp.

Phaeodactylum sp. is a brown-greenish marine diatom belonging to the *Bacillariophyceae*. This microalgae contains high amounts of EPA (Martino et al., 2007). Batista et al. (2017) screened the potential of *Phaeodactylum* sp. incorporation in biscuits.

4.1.4.1. n-3 LC-PUFA and Processing

Although n-3 LC-PUFA rich *Phaeodactylum* was incorporated, no efforts were made to obtain data for the n-3 LC-PUFA incorporation of biscuits with 2 and 6% added biomass.

4.1.4.2. Lipid Oxidation

No lipid oxidation parameters were followed in the study neither.

4.1.4.3. Sensorial Characteristics, Texture and Color

In this study, the texture of biscuits enriched with 2 and 6% *Phaeodactylum* biomass was investigated. The hardness of the biscuits did not differ significantly between the biscuits with 2% addition and the control, while the hardness was significantly higher for the biscuits with 6% addition. The hardness of each biscuit did not change significantly during 8 weeks of storage (Batista et al., 2017).

Lower L* (50 and 35) and a* (-4 and -3) and higher b* (35 and 30) values were detected for biscuits enriched with 2 and 6% biomass, respectively, compared to the control (80, -1 and 25). This can be related to the high amounts of the carotenoid fucoxanthin present in the *Phaeodactylum* biomass. A total color difference (ΔE*) of 7.63, 18.97 and 23.63 for the baked control biscuit and those enriched with 2 and 6% biomass, respectively, was observed compared to the raw control. During an 8 week storage period, the total color differences within a sample were lower than 5, which is not visible by the human eye (Batista et al., 2017).

4.1.5. *Tetraselmis* sp.

Tetraselmis sp. is a green microalga belonging to the *Chlorophyta*. Mainly EPA as n-3 LC-PUFA source can be found in this microalgal species. The total amount of n-3 LC-PUFA is rather limited, which can explain the limited amount of studies focusing on *Tetraselmis* as n-3 LC-PUFA source. The potential use of *Tetraselmis* sp. was screened in bread by Garcia-Segovia et al. (2017).

4.1.5.1. n-3 LC-PUFA and Processing

Garcia-Segovia et al. (2017) incorporated 0.4% *Tetraselmis* in bread, but no further results on the n-3 LC-PUFA incorporation were shown.

4.1.5.2. Lipid Oxidation

No lipid oxidation parameters were followed in this study neither.

4.1.5.3. Sensorial Characteristics, Texture and Color

Garcia-Segovia et al. (2017) screened the influence of 0.4% *Tetraselmis* biomass incorporation on the texture and color parameters of bread. The texture was not significantly modified by the added microalgal biomass.

On the contrary, incorporation of *Tetraselmis* did influence the color. A decrease in the lightness (L* value) from 73 to 62 was observed compared to the control. The redness (a* value) increased from 2.72 to 3.08 compared to the control, while the yellowness (b* value) remained unmodified (Garcia-Segovia et al., 2017).

4.1.6. General Conclusion

The potential of different photoautotrophic microalgae (*Diacronema, Isochrysis, Nannochloropsis, Phaeodactylum* and *Tetraselmis*) has already been screened in food products (Gouveia et al., 2008a; Gouveia et al., 2008b; Fradique et al., 2013; Babuskin et al., 2014; Babuskin et al., 2015; Batista et al., 2017; Garcia-Segovia et al., 2017). However, the range of food products is mostly limited to biscuits and other cereal products. Therefore, it would be interesting to gain more insight into a broader range of food products. All studies described above have used photoautotrophic microalgae as an alternative source of n-3 LC-PUFA, but some of them have not measured the n-3 LC-PUFA concentration of the enriched food product. Those who did, showed that the addition of microalgae is indeed an effective way to increase the n-3 LC-PUFA concentration in the food products. Incorporation concentrations up to 3% of microalgal biomass (*Diacronema, Isochrysis, Nannochloropsis* and *Phaeodactylum)* were in most cases sufficient to claim the products as a source of or rich in n-3 LC-PUFA. Processing may have an impact on the final n-3 LC-PUFA concentration given mechanical and thermal process steps are often exposed at high pressure and temperature, but unfortunately this has rarely been taken into account. Gouveia et al. (2008a and b) and Fradique et al. (2013) found no significant impact of food processing steps (boiling in water or oven baking) on the n-3 LC-PUFA concentration.

The oxidative stability is also an important quality parameter of the final product given the proneness of n-3 LC-PUFA to oxidation, but only Babuskin et al. (2015) has paid attention to it. Both, incorporation of *Isochrysis* and *Nannochloropsis* in chikkis resulted in an increase of the peroxide value with 20 meq active oxygen/kg during 2 months of storage at 27°C. These values are still reasonable low compared to literature, but the authors observed that the addition of supplemental antioxidant may even increase the oxidative stability.

The sensorial characteristics of the enriched food products were marked lower than the control in all studies, but were in most cases still classified positively (Fradique et al., 2013; Babuskin et al., 2014; Babuskin et al., 2015). At higher concentrations (2% in biscuits and 3% in pasta and chikkis) a fishy taste was observed by the panelists, which explains the lower global appreciation of these food products. Similar results for the sensorial characteristics were found for the different microalgal species and the different food products. The firmness of cookies and pasta increased with a higher incorporation concentration, while the texture of chikkis and bread was not affected by the incorporation concentration. Again, no impact of the used species could be detected. Some authors followed textural changes over time, whereby in all cases comparable results were obtained as for the control. The color of the food products changed by incorporation of the pigment rich photoautotrophic microalgae. In general the lightness value L* and the redness value a* decreased, while the yellowness b* value was only limitedly influenced. Studies following the color of the enriched products over time showed no changes (Gouveia et al., 2008b; Babuskin et al., 2014; Babuskin et al., 2015; Batista et al., 2017).

4.2. Heterotrophic Microalgae

4.2.1. *Crypthecodinium* sp.

The microalga *Crypthecodinium* sp. is a heterotrophic microalgae belonging to the *Dinophyceae* and is rich in the fatty acid DHA (Sijtsma & de Swaaf, 2004). The incorporation of *Crypthecodinium* oil was screened in dairy, egg sticks (liquid egg products uniformly cooked using a microwave oven to gel products) and fish and meat products (Park et al., 2004; Chee et al., 2005; Gallaher et al., 2005; Lee et al., 2006; Lopez-Lopez et al., 2009; García-Íñiguez de Ciriano et al., 2010; Kassis et al., 2010; Pietrowski et al., 2011; Sedoski et al., 2012; Alejandre et al., 2017).

4.2.1.1. n-3 LC-PUFA and Processing

A variety of food products were enriched with 0.1 to 10% of *Crypthecodinium* oil, resulting in n-3 LC-PUFA concentrations of 24.4 to 4000 mg/100 g products. The highest amounts are far above the minimum required for an EU nutrition claim (European Commission, 2010). Park et al. (2004) and Pietrowski et al. (2011) enriched surimi with 588 mg n-3 LC-PUFA/100 g. The incorporation concentration in yogurt (0.18% addition) resulted in 147 mg/100 g, while for milk (2-10% addition) amounts between 800 and 4000 mg/100 g were obtained (Gallaher et al., 2005; Lee et al., 2006). Kassis et al. (2010) and Sedoski et al. (2012) enriched egg sticks with 10% heterotrophic microalgal oil to obtain 4000 mg/100 g egg product. Lee et al. (2006), Lopez-Lopez et al. (2009), García-Íñiguez de Ciriano et al. (2010) and Alejandre et al. (2017) incorporated *Crypthecodinium* oil in patties, sausages and ham. The patties were enriched with 24.4 up to 454 mg/100 g product, while the sausages contained 400 up to 3500 mg/100 g final product and ham 454 mg/100 g product. As expected, higher n-3 LC-PUFA concentrations were obtained with higher incorporation concentrations.

Only Lee et al. (2006) and Alejandre et al. (2017) studied the impact of processing. Lee et al. (2006) observed losses of n-3 LC-PUFA up to 20% due to cooking the patties, sausages or ham, while in the study of Alejandre et al. (2017) the cooking process did not affect the n-3 LC-PUFA concentration in the patties. This may suggest the different cooking temperature and time during processing as a possible explanation for the observed differences.

4.2.1.2. Lipid Oxidation

The oxidative stability of surimi was investigated by Park et al. (2004) and Pietrowski et al. (2011). Park et al. (2004) observed an impact of the delivery system of the microalgal n-3 LC-PUFA. During 27 days storage at 6°C, lipid oxidation increased twice as much when the n-3 LC-PUFA was added as emulsified algal oil compared to bulk oil, in both cases without the addition of antioxidants. Addition of cryoprotectants or antioxidants reduced the lipid oxidation, whereby no differences between the enriched samples themselves as well as with the control could be observed anymore. Pietrowski et al. (2011) determined TBARS values (as a marker for secondary oxidation products) after preparing the surimi gels, but did not store the samples longer to follow up further oxidation. An increase of the TBARS value from 0.6 to 0.8 mg MDA/kg sample was observed by the incorporation of *Crypthecodinium* oil compared to the control. Based on both studies it seems that surimi gel enriched with *Crypthecodinium* oil was not oxidatively stable without the addition of antioxidants. Gallaher et al. (2005) and Chee et al. (2005) investigated oxidation in enriched dairy products. Gallaher et al. (2005) could not detect an increase of hydroperoxides (as a marker for primary oxidation) nor propanal (as a marker for secondary oxidation) in any of the incorporation concentrations, while water supplemented with the same amount of *Crypthecodinium* oil started to oxidize. The authors ascribed the antioxidant properties of milk to the casein. It has to be remarked however that 24 hours at 32°C is a short storage period to draw conclusions on oxidative stability. In contrast, hydroperoxides were formed in n-3 LC-PUFA enriched strawberry flavored yogurt during 21 days of storage at 4°C (Chee et al., 2005). Logically, the control without large amounts of unsaturated fatty acids did not oxidize. Based on the results, the authors suggested the need for the addition of antioxidants in enriched food products although the total amount of lipid hydroperoxides was limited to 3 meq/kg after 21 days of storage. Lee et al. (2006) determined the oxidative stability of n-3 LC-PUFA incorporated ham with and without added antioxidants during storage for 21 days under vacuum at -18 and 4°C. A limited increase in the peroxide value from day zero (non-detectable) to day 21 (7.11 and 5.22 µmole/kg), when stored at -18 and 4°C respectively, could be detected in samples without added antioxidants. However, adding a cocktail of antioxidants prevented the products form oxidation. García-Íñiguez de Ciriano et al. (2010) evaluated the oxidative stability of sausages during 30 days of ripening at an unspecified temperature. The TBARS values of the enriched samples as well as of the control were very low (below 0.15 mg MDA/kg) with no significant changes over time, but again antioxidants had been added to control potential oxidation. Alejandre et al. (2017) stored enriched patties vacuum and non-vacuum at 4 and 25°C for 31 days. Vacuum packed patties at 4 and 25°C and non-vacuum packed patties at 4°C remained stable during storage. Patties packed non-vacuum and stored at 25°C showed a significant increase in TBARS (from 0.33 to 0.81 mg MDA/100 g gel). These results again showed that it could be interesting to add some extra antioxidants in order to control lipid oxidation. TBARS values of egg sticks enriched with 4000 mg n-3 LC-PUFA/100 g were not influenced during a 14-days storage at 4°C. This result was however not surprising since the egg sticks were stored under vacuum conditions (Sedoski et al., 2012). It can be concluded that most studies obtained similar results: when storing under non-vacuum conditions additional antioxidants should be added to prevent oxidation.

4.2.1.3. Sensorial Characteristics, Texture and Color

Similar sensorial characteristics were described in different studies (Gallaher et al., 2005; Chee et al., 2005; Lee et al., 2006; García-Íñiguez de Ciriano et al., 2010; Alejandro et al., 2017; Sedoski et al., 2012). For both dairy products, milk and yogurt, the panelists were not able to observe any differences between the control and the enriched food products at day zero (Gallaher et al., 2005; Chee et al., 2005). However, after 15 and 21 days of storage, the sensory panel observed a fishy taste in the enriched products. Enriched meat products, except for hams, were not significantly modified compared to their control (Lee et al., 2006; García-Íñiguez de Ciriano et al., 2010; Alejandre et al., 2017). The enriched hams were marked with 6.4 on a 9-point scale instead of 6.9 for the control (Lee et al., 2006). Furthermore, 55% of the consumer gave a positive answer to the question if they would consume the enriched products (Alejandre et al., 2017). The n-3 LC-PUFA enriched egg sticks (6.0 on a 9-point scale) were accepted by the consumers to the same extent as the control egg sticks (5.9) (Sedoski et al., 2012).

Park et al. (2004) observed that no differences in gel strength were observed between the control (176 g*cm) and surimi gel enriched with emulsified algal oil, while incorporation of bulk oil resulted in a decrease of the gel strength (116 g*cm). On the contrary, Pietrowski et al. (2011)

observed no differences in springiness, cohesiveness, chewiness and resilience compared to the control. The texture of the control egg sticks was more favorably marked (6.3 on a 9-point scale) compared to the enriched sticks (5.2) (Sedoski et al., 2012).

The color of the egg sticks in the study of Sedoski et al. (2012) was not significantly modified compared to the control, while in surimi the b* value increased compared to the control (from 5 to 12 and from 3.72 to 27 respectively) in the studies of Park et al. (2004) and Pietrowski et al. (2011). However in these studies the L* and a* values were unaffected as well. In hams, the L* value increased slightly compared to the control (from 65 to 70), while the a* value was unaffected and no results about the b* value were described (Lee et al., 2006). On the other hand, the L*, a* and b* values were unmodified in fermented sausages (García-Íñiguez de Ciriano et al., 2010).

4.2.2. *Schizochytrium* sp.

The microalga *Schizochytrium* sp. is a heterotrophic microalga belonging to the *Labyrinthulomycetes* and is rich in the fatty acid DHA (Sijtsma & de Swaaf, 2004). The incorporation of *Schizochytrium* was screened in bread, dairy, egg sticks, fish and meat products (Serna-Saldivar et al., 2006; Valencia et al., 2007; Tolasa et al., 2010; Berasategi et al., 2014; Lamas et al., 2015).

4.2.2.1. n-3 LC-PUFA and Processing

Different products were enriched with low amounts (0.325 to 1.32%) of *Schizochytrium* oil, which did however result in relevant amounts of n-3 LC-PUFA in the final product (78-450 mg/100 g product) (Serna-Saldivar et al., 2006; Valencia et al., 2007; Tolasa et al., 2010; Lamas et al., 2015). Only Valencia et al. (2007) used amounts up to 1490 mg/100 g product, which is far above the amount needed for the food claim 'rich in n-3 LC-PUFA'. Obviously, higher incorporation concentrations resulted in a linearly higher n-3 LC-PUFA concentration in the final product, independent of the type of product (Serna-Saldivar et al., 2006; Valencia et al., 2007; Tolasa et al., 2010). None of the studies has taken the impact of processing into account.

4.2.2.2. Lipid Oxidation

Valencia et al. (2007) stored enriched fermented sausages for 90 days at 4°C vacuum as well as non-vacuum packed, but added antioxidants (BHT or BHA). Berasategi et al. (2014) stored enriched sausages for 32 days at 4°C. Again, antioxidants (*M. officinalis* extract) were added. In both studies no signs of oxidation was observed in the enriched products. Tolasa et al. (2010) observed an impact of the quality of the surimi on the oxidative stability, during 4 months of storage at -18 and 3°C. Remarkable is that all samples were stored under vacuum, which already helps to prevent oxidation, but that even under vacuum conditions low quality surimi started to oxidize. The peroxide value increased from 2.5 to 52 meq/kg (from zero to 4 months), but no significant impact of the n-3 LC-PUFA concentration was observed. The peroxide value of high quality surimi on the other hand only increased from 2.5 to 5.3 meq/kg. Frozen storage (-18°C) prevented oxidation during 4 months of storage. Lamas et al. (2015) stored egg sticks enriched with *Schizochytrium* oil for 30 days under frozen conditions. Comparable with the results of Tolasa et al. (2010) no oxidation and no differences compared to the control could be observed.

4.2.2.3. Sensorial Characteristics, Texture and Color

The panelists marked the overall acceptability and flavor of bread enriched with *Schizochytrium* oil the same as the control (2 on a scale between -4 and 4), at day 1 as well as day 13 (Serna-Saldivar et al., 2006). Also enriched egg sticks and sausages obtained the same results as the control by a consumer panel (Berasategi et al., 2014; Lamas et al., 2015). Valencia et al. (2007) found similar result for fermented sausages enriched with 15% *Schizochytrium* oil, although the panelists indicated the taste in a triangle test significantly different for sausages enriched with 25% *Schizochytrium* oil compared to the control.

No significant differences in texture were found in enriched bread and sausages compared to the control (Serna-Saldivar et al., 2006; Berasategi et al., 2014). However, the texture of surimi was influenced by the incorporation concentration as well as the quality of the gel. High quality surimi with an incorporation concentration of 0.5% had the highest compressive force (62.66 kg), followed by high quality surimi with 1% microalgal oil (29.50 kg) and low quality surimi with 0.5 and 1% microalgal oil (8 kg) (Tolasa et al., 2010).

Color was only determined in the studies of Berasategi et al. (2014) and Lamas et al. (2015). The L*, a* and b* values were significantly higher in the control egg sticks (66.5, 22 and 62.5) compared to the enriched egg sticks (65, 19 and 50.8), while for sausages the opposite was observed. The enriched sausages had similar a* values (21) as the control, but higher L* (64 instead of 60) and b* (9 instead of 5) values. The differences can be explained by the fact that in the egg sticks not only *Schizochytrium* oil was added, but the egg yolk was removed as well.

4.2.3. General Conclusion

The potential of different heterotrophic oils (*Crypthecodinium* and *Schizochytrium*) was screened in bread, egg sticks, surimi, patties, ham and sausages (Park et al., 2004; Chee et al., 2005; Gallaher et al., 2005; Lee et al., 2006; Serna-Saldivar et al., 2006; Valencia et al., 2007; Lopez-Lopez et al., 2009; García-Íñiguez de Ciriano et al., 2010; Kassis et al., 2010; Tolasa et al., 2010; Pietrowski et al., 2011; Sedoski et al., 2012; Berasategi et al., 2014; Lamas et al., 2015; Alejandre et al., 2017). The oil of both species is authorized by the EFSA and FDA (EFSA, 2011; FDA, 2014). A wide variety of incorporation concentrations (0.1 up to 10%) was used in the studies, some of them far above the amount needed for the food claim 'rich in n-3 LC-PUFA'. Oil from heterotrophic sources allows to reach high amounts of n-3 LC-PUFA in the final product with low incorporation concentrations. Obviously, higher incorporation concentrations resulted in a linearly higher n-3 LC-PUFA concentration in the final product, independently of the type of product (Serna-Saldivar et al., 2006; Valencia et al., 2007; Tolasa et al., 2010). Following the oxidative stability of food products enriched with heterotrophic oil was often executed under vacuum and by preventively adding of antioxidants. However, even under these conditions, it was not always possible to keep the enriched products oxidatively stable showing the need for preventive actions. Sensorial characteristics and texture were only seldom changed compared to the control by the incorporation of heterotrophic oil. The color was only limitedly influenced and mainly in the b* value, which expresses changes in yellowness.

5. Comparison Photoautotrophic and Heterotrophic Microalgae

Both photoautotrophic and heterotrophic microalgae have shown their potential as an alternative source of n-3 LC-PUFA in different food products with each their advantages and disadvantages. Photoautotrophic biomass showed a higher oxidative stability than heterotrophic oil. With incorporation of heterotrophic microalgae demanding the addition of antioxidants or vacuum packaging, which was even not always enough to maintain the oxidative stability. On the other hand, the presence of pigments in photoautotrophic species resulted in larger changes in color of the enriched product compared to the heterotrophic species. The controls showed similar sensorial characteristics as products enriched with heterotrophic microalgal oil, while the control was preferred over products enriched with photoautotrophic biomass although the latter were still positively classified. However, the incorporation concentration was limited to a maximum (depending on food product and species). Above this limit, a fishy taste was observed. The texture was mainly influenced by the addition of photoautotrophic biomass and was logically depending on the concentration.

Addition of microalgae as a source of n-3 LC-PUFA to food products is thus an effective way to increase n-3 LC-PUFA levels in the diet. However, comparison of the different studies was not always straight forward due to differences in set-up (food products, concentrations, storage conditions) and none of the studies used photoautotrophic as well as heterotrophic microalgae in the same study. Therefore, further research on the potential of microalgae as n-3 LC-PUFA source in a broader range of food products, in standardized set ups, paying attention to the impact of processing, … and development in the legislation to authorize new species, should be executed to support prospective applications.

ACKNOWLEDGMENT

This work was financially supported by the Research Foundation - Flanders (FWO SB PhD fellowship L. Gheysen 1S27016N), Industrial Research Fund KULeuven (IOF-KP Vegetalgae - KP/14/004) and KULeuven Kulak (Valerie Matton).

REFERENCES

Adarme-Vega, T., Lim, D. K. Y., Timmins, M., Vernen, F., Li, Y., Schenk, P. M. (2012). Microalgal biofactories: a promising approach towards sustainable omega-3 fatty acid production. *Microbial Cell Factories*, *11*(1), 96.

Alejandre, M., Passarini, D., Astiasarán, I., Ansorena, D. (2017). The effect of low-fat beef patties formulated with a low-energy fat analogue enriched in long-chain polyunsaturated fatty acids on lipid oxidation and sensory attributes. *Meat Science*, *134*, 7–13.

Anbudhasan, P., Surendraraj, A., Karkuzhali, S., Ramasamy, D. (2014). Development of Omega 3 Fatty Acid Enriched Stable Functional Foods: Challenges and Options. *International Journal of Innovative Research and Review*, *2*(1), 1–13.

Babuskin, S., Krishnan, K. R., Babu, P. A. S., Sivarajan, M., Sukumar, M. (2014). Functional foods enriched with marine microalga

Nannochloropsis oculata as a source of omega-3 fatty acids. *Food Technology and Biotechnology*, *52*(3), 292–299.

Babuskin, S., Radhakrishnan, K., Azhagu Saravana Babu, P., Sukumar, M., Fayidh, M. A., Sabina, K., Archana, G., Sivarajan, M. (2015). Effects of Rosemary extracts on oxidative stability of chikkis fortified with microalgae biomass. *Journal of Food Science and Technology*, *52*(6), 3784–3793.

Balduyck, L., Bijttebier, S., Bruneel, C., Jacobs, G., Voorspoels, S., Van Durme, J., Muylaert, K., Foubert, I. (2016). Lipolysis in T-Isochrysis lutea during wet storage at different temperatures. *Algal Research*, *18*, 281–287.

Balduyck, L., Stock, T., Bijttebier, S., Bruneel, C., Voorspoels, S., Muylaert, K., Foubert, I. (2017). Integrity of the microalgal cell wall plays a major role in the lipolytic stability during wet storage. *Algal Research*, *25*, 1–30.

Batista, A. P., Niccolai, A., Fradinho, P., Fragoso, S., Bursic, I., Rodolfi, L., Biondi, N., Tredici, M., Sousa, I., Raymundo, A. (2017). Microalgae biomass as an alternative ingredient in cookies: Sensory, physical and chemical properties, antioxidant activity and in vitro digestibility. *Algal Research*, *26*, 161–171.

Bellou, S., Triantaphyllidou, I. E., Aggeli, D., Elazzazy, A. M., Baeshen, M. N., Aggelis, G. (2016). Microbial oils as food additives: Recent approaches for improving microbial oil production and its polyunsaturated fatty acid content. *Current Opinion in Biotechnology*, *37*, 24–35.

Bendif, E. M., Probert, I., Hervé, A., Billard, C., Goux, D., Lelong, C., Cadoret, J. P., Véron, B. (2011). Integrative Taxonomy of the Pavlovophyceae (Haptophyta): A Reassessment. *Protist*, *162*(5), 738–761.

Bendif, E. M., Probert, I., Schroeder, D. C., de Vargas, C. (2013). On the description of Tisochrysis lutea gen. nov. sp. nov. and Isochrysis nuda sp. nov. in the Isochrysidales, and the transfer of Dicrateria to the Prymnesiales (Haptophyta). *Journal of Applied Phycology*, *25*(6), 1763–1776.

Berasategi, I., Navarro-Blasco, Í., Calvo, M. I., Cavero, R. Y., Astiasarán, I., Ansorena, D. (2014). Healthy reduced-fat Bologna sausages enriched in ALA and DHA and stabilized with Melissa officinalis extract. *Meat Science*, *96*(3), 1185–1190.

Boelen, P., van Mastrigt, A., van de Bovenkamp, H. H., Heeres, H. J., Buma, A. G. J. (2017). Growth phase significantly decreases the DHA-to-EPA ratio in marine microalgae. *Aquaculture International*, *25*(2), 577–587.

Borowitzka, M. A., Moheimani, N. R. (2013). *Algae for Biofuels and Energy*. Springer, 288 p.

Brasseur, D., Delzenne, N., Henderickx, H., Huyghebaert, A., Kornitzer, M., Ulens, M. (2004). Aanbevelingen en beweringen betreffende de Omega-3 vetzuren, versie 2004. [Recommendations and claims concerning Omega-3 fatty acids, version 2004] *Hoge Gezondheidsraad*, 108 p.

Bumbak, F., Cook, S., Zachleder, V., Hauser, S., Kovar, K. (2011). Best practices in heterotrophic high-cell-density microalgal processes: Achievements, potential and possible limitations. *Applied Microbiology and Biotechnology*, *91*(1), 31–46.

Buono, S., Langellotti, Antonio Luca Martello, A., Rina, F., Fogliano, V. (2014). Functional ingredients from microalgae. *Functional Food*, 1669–1685.

Calder, P. C. (2014). Very long chain omega-3 (n-3) fatty acids and human health. *European Journal of Lipid Science and Technology*, *116*(10), 1–21.

Chacón-Lee, T. L., González-Mariño, G. E. (2010). Microalgae for "Healthy" Foods-Possibilities and Challenges. *Comprehensive Reviews in Food Science and Food Safety*, *9*(6), 655–675.

Chee, C. P., Gallaher, J. J., Djordjevic, D., Faraji, H., McClements, D. J., Decker, E., Hollender, R., Peterson, D., Roberts, R., Coupland, J. N. (2005). Chemical and sensory analysis of strawberry flavoured yogurt supplemented with an algae oil emulsion. *The Journal of Dairy Research*, *72*(3), 311–316.

Chen, H. H., Jiang, J. G. (2017). Lipid Accumulation Mechanisms in Auto- and Heterotrophic Microalgae. *Journal of Agricultural and Food Chemistry*, *65*(37), 8099–8110.

Chen, M., Tang, H., Ma, H., Holland, T. C., Ng, K. Y. S., Salley, S. O. (2011). Effect of nutrients on growth and lipid accumulation in the green algae Dunaliella tertiolecta. *Bioresource Technology*, *102*(2), 1649–1655.

Covington, M. B. (2004). Omega-3 fatty acids. *American Family Physician*, 70, 133-140.

Deckelbaum, R. J., Torrejon, C. (2012). The Omega-3 Fatty Acid Nutritional Landscape: Health Benefits and Sources. *Journal of Nutrition*, *142*(3), 587S–591S.

Dineshkumar, R., Narendran, R., Sampathkumar, P. (2017). Cultivation and harvesting of micro-algae for bio-fuel Production – A review. *Indian Journal of Geo Marine Science,* 46, 1731–1742.

EFSA. (2011). Scientific Opinion on the substantiation of health claims related to docosahexaenoic acid (DHA) and eicosapentaenoic acid (EPA). *EFSA Journal*, *9*(1924), 1–30.

European Commission. (2010). Verordening (EU) Nr. 116/2010, (116), 2009–2011.

Fawley, M. W., Jameson, I., Fawley, K. P. (2015). New insights into the morphology and phylogeny of Heterosigma akashiwo (Raphidophyceae), with the description of Heterosigma minor sp. nov. *Phycologia*, *54*(5), 545–552.

FDA. (2014). GRAS determination of DHA algal Oil for Use in Infant Formula, *GRAS notice No. 667.*

Flock, M. R., Harris, W. S., Kris-Etherton, P. M. (2013). Long-chain omega-3 fatty acids: Time to establish a dietary reference intake. *Nutrition Reviews*, *71*(10), 692–707.

Fradique, M., Batista, A. P., Nunes, M. C., Gouveia, L., Bandarra, N. M., Raymundo, A. (2013). Isochrysis galbana and Diacronema vlkianum biomass incorporation in pasta products as PUFA's source. *LWT - Food Science and Technology*, *50*(1), 312–319.

Fraeye, I., Bruneel, C., Lemahieu, C., Buyse, J., Muylaert, K., Foubert, I. (2012). Dietary enrichment of eggs with omega-3 fatty acids: A review. *Food Research International*, *48*(2), 961–969.

Frankel, N. (1980). Lipid oxidation. *Pro 0. Lipid Res*, *19*, 1–22.

Gallaher, J. J., Hollender, R., Peterson, D. G., Roberts, R. F., Coupland, J. N. (2005). Effect of composition and antioxidants on the oxidative stability of fluid milk supplemented with an algae oil emulsion. *International Dairy Journal*, *15*(4), 333–341.

García-Íñiguez de Ciriano, M., Larequi, E., Rehecho, S., Calvo, M. I., Cavero, R. Y., Navarro-Blasco, Í., Astiasaran, I., Ansorena, D. (2010). Selenium, iodine, ω-3 PUFA and natural antioxidant from Melissa officinalis L.: A combination of components from healthier dry fermented sausages formulation. *Meat Science*, *85*(2), 274–279.

Garcia-Segovia, M., Pagan-Moreno, I., Martinez-Monzo, L. (2017). Effect of microalgae incorporation on physiochemical and textural properties in wheat bread formulation. *Food Science and Technology International*, 23(5), 437-447.

Gnouma, A., Sadovskaya, I., Souissi, A., Sebai, K., Medhioub, A., Grard, T., Souissi, S. (2017). Changes in fatty acids profile, monosaccharide profile and protein content during batch growth of Isochrysis galbana (T.iso). *Aquaculture Research*, *48*(9), 4982–4990.

GOED. (2016). *EPA + DHA daily intake recommendation GOED 700-1000 mg, 10*(7), 201604.

Gouveia, L., Batista, A. P., Raymundo, A., Bandarra, N. (2008a). *Spirulina maxima* and *Diacronema vlkianum* microalgae in vegetable gelled desserts. *Nutrition & Food Science*, *38*(5), 492–501.

Gouveia, L., Coutinho, C., Mendon, E., Batista, A. P., Sousa, I., Bandarra, N. M., Raymundo, A. (2008b). Functional biscuits with PUFA-ω3 from *Isochrysis galbana*. *Journal of the Science of Food and Agriculture*, *88*, 891–896.

Hillebrand, H., Sommer, U. (1999). The nutrient stoichiometry of benthic microalgal growth: Redfield proportions are optimal. *Limnology and Oceanography*, *44*(2), 440–446.

Hu, C., Li, M., Li, J., Zhu, Q., Liu, Z. (2008). Variation of lipid and fatty acid compositions of the marine microalga Pavlova viridis (Prymnesiophyceae) under laboratory and outdoor culture conditions. *World Journal of Microbiology and Biotechnology*, *24*(7), 1209–1214.

Hubert, F., Poisson, L., Loiseau, C., Gauvry, L., Pencréac'h, G., Hérault, J., Ergan, F. (2017). Lipids and lipolytic enzymes of the microalga *Isochrysis galbana*. *Ocl*, *24*(4), D407.

Jasvinder, S., Saxena, R. C. (2015). Chapter 2: An Introduction to Microalgae: Diversity and Significance. In *Handbook of Marine Microalgae Biotechnology Advances* (pp. 11–24).

Kassis, N. M., Beamer, S. K., Matak, K. E., Tou, J. C., Jaczynski, J. (2010). Nutritional composition of novel nutraceutical egg products developed with omega-3-rich oils. *LWT - Food Science and Technology*, *43*(8), 1204–1212.

Kris-Etherton, P. M., Grieger, J. a., Etherton, T. D. (2009). Dietary reference intakes for DHA and EPA. *Prostaglandins Leukotrienes and Essential Fatty Acids*, *81*(2-3), 99–104.

Kris-Etherton, P. M., Taylor, D. S., Yu-Poth, S., Huth, P., Moriarty, K., Fishell, V., Hargrove, R.L., Zhao, G., Etherton, T. D. (2000). Polyunsaturated fatty acids in the food chain in the United States. *The American Journal of Clinical Nutrition*, *71*, 179S–88S.

Lamas, A., X., Miranda, J. M., Roca-Saavedra, P., Cardelle-Cobas, a., Rodriguez, J. a., Franco, C.M., Cepeda, A. (2015). Technological development of functional egg products by an addition of n- 3 polyunsaturated-fatty-acid-enriched oil. *CyTA - Journal of Food*, 14(2), 289-295.

Lee, S., Hernandez, P., Djordjevic, D., Faraji, H., Hollender, R., Faustman, C., Decker, E. (2006). Effect of Antioxidants and Cooking on Stability of n-3 Fatty Acids in Fortified Meat Products. *Journal of Food Science*, *71*(3), 233–238.

Lopez-Lopez, I., Cofrades, S., Jiménez-Colmenero, F. (2009). Low-fat frankfurters enriched with n-3 PUFA and edible seaweed: Effects of olive oil and chilled storage on physicochemical, sensory and microbial characteristics. *Meat Science*, *83*(1), 148–154.

Madeira, M. S., Cardoso, C., Lopes, P. a., Coelho, D., Afonso, C., Bandarra, N. M., Prates, J. a. M. (2017). Microalgae as feed ingredients for livestock production and meat quality: A review. *Livestock Science*, *205*, 111–121.

Martino, A. De, Meichenin, A., Shi, J., Pan, K., Bowler, C. (2007). Genetic and phenotypic characterization of Phaeodactylum tricornutum (Bacillariophyceae) accessions. *Journal of Phycology*, *43*(5), 992–1009.

Matos, J., Cardoso, C., Bandarra, N. M., Afonso, C. (2017). Microalgae as healthy ingredients for functional food: a review. *Food Funct.*, 2672–2685.

Mercer, P., Armenta, R. E. (2011). Developments in oil extraction from microalgae. *European Journal of Lipid Science and Technology*, *113*(5), 539–547.

Milledge, J. J., Heaven, S. (2013). A review of the harvesting of micro-algae for biofuel production. *Reviews in Environmental Science and Biotechnology*, *12*(2), 165–178.

Molendi-Coste, O., Legry, V., Leclercq, I. (2011). Why and how meet n-3 PUFA dietary recommendations? *Gastroenterology Research and Practice*, *2011*, 1–11.

Morales-Sánchez, D., Martinez-Rodriguez, O., Kyndt, J., Martinez, A. (2014). Heterotrophic growth of microalgae: metabolic aspects. *World Journal of Microbiology and Biotechnology*, *31*(1), 1–9.

Nalder, T. D., Miller, M. R., & Packer, M. (2015). Changes in lipid class content and composition of Isochrysis sp. (T-Iso) grown in batch culture. *Aquaculture International*, *23*(5), 1293–1312.

Park, Y., Kelleher, S. D., McClements, D. J., Decker, E. (2004). Incorporation and stabilization of omega-3 fatty acids in surimi made from cod, Gadus morhua. *Journal of Agricultural and Food Chemistry*, *52*(3), 597–601.

Pietrowski, B. N., Tahergorabi, R., Matak, K. E., Tou, J. C., Jaczynski, J. (2011). Chemical properties of surimi seafood nutrified with ω-3 rich oils. *Food Chemistry*, *129*(3), 912–919.

Pignolet, O., Jubeau, S., Vaca-Garcia, C., Michaud, P. (2013). Highly valuable microalgae: Biochemical and topological aspects. *Journal of Industrial Microbiology and Biotechnology*, *40*(8), 781–796.

Pires, J. (2015). Chapter 5: Mass production of microalgae. In *Handbook of Marine Microalgae Biotechnology Advances* (pp. 55–68).

Rocha, J. M. S., Garcia, J. E. C., Henriques, M. H. F. (2003). Growth aspects of the marine microalga Nannochloropsis gaditana. *Biomolecular Engineering*, *20*(4-6), 237–242.

Russo, G. L. (2009). Dietary n - 6 and n - 3 polyunsaturated fatty acids: From biochemistry to clinical implications in cardiovascular prevention. *Biochemical Pharmacology*, *77*(6), 937–946.

Ryckebosch, E., Bruneel, C., Muylaert, K., Foubert, I. (2012). Microalgae as an alternative source of omega-3 long chain polyunsaturated fatty acids. *Lipid Technology*, *24*(6), 128–130.

Ryckebosch, E., Bruneel, C., Termote-Verhalle, R., Goiris, K., Muylaert, K., Foubert, I. (2014). Nutritional evaluation of microalgae oils rich in omega-3 long chain polyunsaturated fatty acids as an alternative for fish oil. *Food Chemistry*, *160*, 393–400.

Ryckebosch, E., Bruneel, C., Termote-Verhalle, R., Lemahieu, C., Muylaert, K., Van Durme, J., Goiris, K., Foubert, I. (2013). Stability of omega-3 LC-PUFA-rich photoautotrophic microalgal oils compared to commercially available omega-3 LC-PUFA oils. *Journal of Agricultural and Food Chemistry*, *61*(42), 10145–55.

Ryckebosch, E., Muylaert, K., Eeckhout, M., Ruyssen, T., Foubert, I. (2011). Influence of drying and storage on lipid and carotenoid stability of the Microalga Phaeodactylum tricornutum. *Journal of Agricultural and Food Chemistry*, *59*(20), 11063–11069.

Salim, S., Kosterink, N. R., Tchetkoua Wacka, N. D., Vermuë, M. H., Wijffels, R. H. (2014). Mechanism behind autoflocculation of unicellular green microalgae Ettlia texensis. *Journal of Biotechnology*, *174*(1), 34–38.

Sedoski, H. D., Beamer, S. K., Jaczynski, J., Partington, S., Matak, K. E. (2012). Sensory evaluation and quality indicators of nutritionally-enhanced egg product with omega-3 rich oils. *LWT - Food Science and Technology*, *47*(2), 459–464.

Serna-Saldivar, S. O., Zorrilla, R., De La Parra, C., Stagnitti, G., Abril, R. (2006). Effect of DHA containing oils and powders on baking performance and quality of white pan bread. *Plant Foods for Human Nutrition*, *61*(3), 121–129.

Sidhu, K. S. (2003). Health benefits and potential risks related to consumption of fish or fish oil. *Regulatory Toxicology and Pharmacology*, *38*(3), 336–344.

Sijtsma, L., de Swaaf, M. E. (2004). Biotechnological production and applications of the omega-3 polyunsaturated fatty acid docosahexaenoic acid. *Applied Microbiology and Biotechnology*, *64*(2), 146–53.

Simopoulos, A. P. (2002). The importance of the ratio of omega-6/omega-3 essential fatty acids. *Biomedicine and Pharmacotherapy*, *56*(8), 365–379.

Sioen, I., Devroe, J., Inghels, D., Terwecoren, R., De Henauw, S. (2010). The influence of n-3 PUFA supplements and n-3 PUFA enriched foods on the n-3 LC PUFA intake of flemish women. *Lipids*, *45*(4), 313–320.

Suali, E., Sarbatly, R. (2012). Conversion of microalgae to biofuel. *Renewable and Sustainable Energy Reviews*, *16*(6), 4316–4342.

Swanson, D., Block, R., Mousa, S. (2012). Omega-3 fatty acids EPA and DHA: health benefits throughout life. *Advances in Nutrition: An Internation Review Journal*, *3*, 1–7.

Tolasa, S., Lee, C. M., Cakli, S. (2010). Physical and Oxidative Stabilization of Omega-3 Fatty Acids in Surimi Gels. *Journal of Food Science*, *75*(3), 305–310.

Valencia, I., Ansorena, D., Astiasarán, I. (2007). Development of dry fermented sausages rich in docosahexaenoic acid with oil from the microalgae Schizochytrium sp.: Influence on nutritional properties, sensorial quality and oxidation stability. *Food Chemistry*, *104*(3), 1087–1096.

Vandamme, D. (2016). CHAPTER II.4 Harvesting, thickening and dewatering processes. In *Microalgae as a Source of Bioenergy: Products, Processes and Economics* (pp. 202–223).

Vandamme, D., Foubert, I., Muylaert, K. (2013). Flocculation as a low-cost method for harvesting microalgae for bulk biomass production. *Trends in Biotechnology*, *31*(4), 233–9.

Walker, R., Decker, E. a., McClements, D. J. (2015). Development of food-grade nanoemulsions and emulsions for delivery of omega-3 fatty acids: opportunities and obstacles in the food industry. *Food Funct.*, *6*(1), 41–54.

Yokoyama, R., Honda, D. (2007). Taxonomic rearrangement of the genus Schizochytrium sensu lato based on morphology, chemotaxonomic characteristics, and 18S rRNA gene phylogeny (Thraustochytriaceae, Labyrinthulomycetes): Emendation for Schizochytrium and erection of Aurantiochytrium and Oblongichytrium gen. nov. *Mycoscience*, 48(4), 199-211.

Zárate, R., el Jaber-Vazdekis, N., Tejera, N., Pérez, J., Rodríguez, C. (2017). Significance of long chain polyunsaturated fatty acids in human health. *Clinical and Translational Medicine*, 6(1), 1-19.

In: Polyunsaturated Fatty Acids (PUFAs) ISBN: 978-1-53613-572-5
Editor: Angel Catala

Chapter 2

POLYUNSATURATED FATTY ACIDS (PUFAS) IN HUMAN HEALTH

Takayuki Ohwada*[*]*, MD
Department of Cardiology, Fukushima Red Cross Hospital,
Fukushima, Japan

ABSTRACT

Polyunsaturated fatty acids (PUFAs), which are known to be critical for human health, are lipids in which the constituent hydrocarbon chain possesses two or more carbon-carbon double bonds. PUFAs are distributed in almost all cells in the human body and affect membrane composition and function, eicosanoid synthesis, signal transduction, and gene expression. Linoleic acid (LA), an omega-6 fatty acid, and α-linolenic acid (ALA), an omega-3 fatty acid, are considered essential fatty acids because they cannot be synthesized by humans. The omega-3 fatty acids eicosapentaenoic acid (EPA) and docosahexaenoic acid (DHA) can be synthesized from ALA; however, due to low conversion efficiency, EPA and DHA should be obtained from food sources. Foods that provide omega-3 fatty acids include fish oil and certain plant and nut oils, whereas omega-6 fatty acids can be found in palm, soybean, rapeseed,

[*] Corresponding Author Email: ikyoku18@fukushima-med-jrc.jp.

and sunflower oils. Omega-9 fatty acids, which are not considered essential, can be found in animal fat and olive oil. Fish oil contains two omega-3 fatty acids, i.e., DHA and EPA. Some nuts, seeds, and vegetable oils contain ALA, which may be converted into DHA and EPA in the body. Importantly, omega-6 PUFAs have been shown to promote inflammation and augment many diseases, whereas omega-3 PUFAs appear to counter these adverse effects. Indeed, omega-3 fatty acids are thought to provide a wide range of health benefits, including reducing the risk of coronary heart disease and improving cholesterol levels. Studies of omega-3 fatty acids have shown promising effects in cancer, diabetes mellitus, depression, and attention-deficit hyperactivity disorder. Due to these potential health benefits, fish oil, which is rich in omega-3 fatty acids, has become a popular supplement. Moreover, recommended amounts of DHA and EPA, consumed from fish or fish oil supplements, may lower triglycerides and reduce the risk of heart attack, arrhythmia, and stroke in individuals with heart disorders.

To induce biological effects, PUFAs must be metabolized to their respective long-chain metabolites. PUFAs not only form precursors to respective prostaglandins (PGs), thromboxanes (TXs), and leukotrienes (LTs) but also modify lipoxins (LXs), resolvins, and isoprostanes. Certain PGs, TXs, and LTs have pro-inflammatory effects, whereas LXs and resolvins are anti-inflammatory molecules. Furthermore, PUFAs and their long-chain metabolites modulate the activities of angiotensin-converting enzymes (ACE) and HMG-CoA reductase enzymes, enhance acetylcholine levels in the brain, increase the synthesis of endothelial nitric oxide, augment diuresis, and enhance insulin action. Thus, PUFAs and their metabolites may function as endogenous ACE and HMG-CoA reductase inhibitors, nitric oxide enhancers, β-blockers, diuretics, antihypertensive, and anti-atherosclerotic molecules. Additionally, PUFAs and their long-chain metabolites affect signal transduction via activation of peroxisome proliferator-activated receptors. Thus, PUFAs and their derivatives are involved in many physiological and pathological actions to regain homeostasis.

Keywords: Polyunsaturated Fatty Acids, Cardiovascular Disease, Inflammatory Bowel Disease, Alzheimer Disease, Parkinson Disease

1. CELL

Polyunsaturated fatty acids (PUFA) form the lipid bilayer of the cell membranes and make a key role in the cellular adaptation to the environment. They affect cell membrane properties, such as fluidity, flexibility, permeability, and the activity of membrane-bound enzymes [1]. PUFA linolenic acid prevents neuronal death in an animal model of transient global ischemia even when administered after the insult [2]. Linolenic acid also protects animals treated with kainate against seizures and hippocampal lesions. The same effects have been observed in an in vitro model of seizure-like activity using glutamatergic neurons and they have been shown to be associated with blockade of glutamatergic transmission by low concentrations of distinct polyunsaturated fatty acids. The opening of background K (+) channels, like TREK-1 and TRAAK, which are activated by arachidonic acid and other polyunsaturated fatty acids such as docosahexaenoic acid and linolenic acid, is a significant factor in this neuroprotective effect. These channels are abundant in the brain where they are located both pre- and post-synaptically and are insensitive to saturated fatty acids, which offer no neuroprotection. Dietary n-3 PUFAs have been reported to exhibit antiarrhythmic properties, and these effects have been attributed to their capability to modulate ion channels. N-3 PUFAs of marine (docosahexaenoic, DHA and eicosapentaenoic acid, EPA) and plant origin (alpha-linolenic acid, ALA) block K (v) 1.5 and K (v) 11.1 channels at physiological concentrations [3]. Moreover, DHA and EPA decrease the expression levels of K (v) 1.5, whereas ALA does not. DHA and EPA also decrease the magnitude of the currents elicited by the activation of Na (v) 1.5 and calcium channels. These effects on sodium and calcium channels should theoretically shorten the cardiac action potential duration (APD), whereas the blocking actions of n-3 PUFAs on K (v) channels would be expected to produce a lengthening of the cardiac action potential.

2. Vision

The long-term effect on the visual development of omega-3 polyunsaturated fatty acid (n-3 PUFA) intake during gestation was reported [4]. Using visual evoked potentials (VEPs), the long-term effects on visual development were evaluated in 136 school-age Inuit children exposed to high levels of n-3 PUFAs during gestation. VEP protocols using color and motion stimuli were used to assess parvocellular and magnocellular responses. Concentrations of the two-major n-3 PUFAs (DHA and EPA) were measured in the umbilical cord and child plasma phospholipids, reflecting prenatal and postnatal exposure, respectively. Cord plasma DHA level was found to be associated with shorter latencies of the N1 and P1 components of the color VEPs. This study demonstrates beneficial effects of DHA intake during gestation on visual system function at school age. DHA is particularly important for the early development and long-term function of the visual parvocellular pathway.

Accumulation and aggregation of misfolded proteins is a hallmark of several diseases collectively known as proteinopathies. Autophagy has a cytoprotective role in diseases associated with protein aggregates [5]. Age-related macular degeneration (AMD) is the most common neurodegenerative eye disease that evokes blindness in elderly. AMD is characterized by degeneration of retinal pigment epithelial (RPE) cells and leads to loss of photoreceptor cells and central vision. The initial phase associated with accumulation of intracellular lipofuscin and extracellular deposits called drusen. Epidemiological studies have suggested an inverse correlation between dietary intake of marine n-3 PUFAs and the risk of developing neurodegenerative diseases, including AMD. However, the disease-preventive mechanism(s) mobilized by n-3 PUFAs is not completely understood. In human retinal pigment epithelial cells, Johansson I et al. found that physiologically relevant doses of the N-3 PUFA DHA induce a transient increase in cellular reactive oxygen species (ROS) levels that activates the oxidative stress response regulator NFE2L2/NRF2 (nuclear factor, erythroid derived 2, like 2). Simultaneously, there is a transient increase in intracellular protein aggregates containing SQSTM1/p62 (sequestosome 1) and an increase in autophagy. Pretreatment with DHA rescues the cells from cell cycle arrest induced by misfolded proteins or oxidative stress. Cells with a downregulated oxidative stress response, or autophagy, respond with reduced cell growth and survival after DHA supplementation. These results suggest that DHA both induces endogenous antioxidants and mobilizes selective autophagy of misfolded proteins. Both mechanisms could be relevant to reduce the risk of developing aggregate-associate diseases such as AMD. Whereas, in another study [6], dietary oily fish and seafood intake were significantly lower in AMD patients than in controls. Serum EPA was associated significantly with a lower risk for neovascular AMD (odds ratio [OR] = 0.41; 95% confidence interval [CI], 0.22-0.77; P = 0.005). Analysis of red blood cell membranes revealed that EPA and EPA+DHA were associated significantly with a lower risk for neovascular AMD (OR = 0.25; 95%CI, 0.13-0.47; P<0.0001 and OR = 0.52; 95% CI, 0.29-0.94; P = 0.03, respectively).

To assess the associations between dietary intake of PUFAs and diabetic retinopathy (DR), 379 patients (median age: 66.0 years) with diabetic retinopathy attending a diabetes eye clinic were categorized as "well-controlled diabetes" (n = 123) and "poorly controlled diabetes" (n = 256), defined as glycated hemoglobin (HbA1c) level < 7.0% or ≥ 7.0%, respectively [7]. Among patients with well-controlled diabetes, increasing daily intake of PUFAs was associated with a reduced likelihood of the presence (OR: 0.18; 95% CI: 0.06-0.59) and severity of DR after adjusting for age, sex, HbA1c, mean arterial blood pressure, and duration of diabetes. Moreover, an increased saturated fatty acid (SFA) intake was associated with increased likelihood of the presence (OR: 2.37; 95% CI: 1.15-4.88) and severity of DR. No association was found among those with poorly controlled diabetes.

3. Cognitive Function

In the last decade, the association between diet and cognitive function or dementia has been largely investigated. Some cohort studies have shown an inverse association between fish consumption and the prevalence of dementia or Alzheimer disease (AD) [8-11], but these findings were not supported by another study [12]. Meta-analyses have also reported conflicting findings [13-16]. There is still a negative correlation of cognitive functions with saturated fatty acids and a protective effect against the cognitive decline of elevated fish consumption, high intake of monounsaturated fatty acids and PUFA, particularly n-3 PUFA [17]. Moreover, Zang et al. [18] demonstrated by a dose-response meta-analysis of 21cohort studies that 0.1 g/d increment of dietary DHA intake was associated with lower risks of dementia (RR: 0.86;95%CI: 0.76-0.96; P < 0.001, I (2) = 92.7%) and AD (RR: 0.63; 95% CI: 0.51-0.76; P < 0.001, I (2) = 94.5%). Significant curvilinear relations between fish consumption and risk of the AD and between total PUFAs and risk of mild cognitive impairment (MCI) (both P-nonlinearity < 0.001) were observed. Chronic stress induces dendritic atrophy in the hippocampus and impairs learning and memory, supplementation with n-3 PUFA is known to improve learning and memory of control rats. Perez et al. [19] demonstrated that supplementation with n-3 PUFA improved dendritic architecture and restored the frequency of inhibitory post-synaptic currents of hippocampal pyramidal neurons of rats receiving stress. In addition, n-3 PUFA supplementation improved spatial memory. N-3 PUFA supplementation had three beneficial effects on stressed rats: prevented or compensated dendritic atrophy in CA3; restored the probability of GABA release in CA1; and improved spatial memory. They argued that n-3 PUFA supplementation can be used in treating stress-related psychiatric disorders such as depression and anxiety.

4. Cardioprotective Function

Long-chain polyunsaturated fatty acids (LC-PUFAs) have cardioprotective functions, i.e., reduction of platelet aggregation and inflammation, antiarrhythmic effect, improvement of endothelial function and myocardial relaxation and efficiency, stabilization of atherosclerotic plaques, and reduction of resting heart rate and systolic and diastolic blood pressure. However, so far, there are no reports from randomized controlled trials (RCTs) that have targeted exclusively the primary prevention of coronary heart disease (CHD), that is, the effects of omega-3 PUFA supplements in the general population of patients without prior CHD. No recommendation for treatment with omega-3 PUFA supplements can be made for this segment of the population.

4.1. PUFA for High-Risk Patients for Cardiovascular Disease (CVD)

In 2002, the American Heart Association (AHA) published a scientific statement, "Fish Consumption, Fish Oil, Omega-3 Fatty Acids, and Cardiovascular Disease" [20]. At that time, evidence from 2 large RCTs demonstrated that EPA+DHA supplements significantly reduced fatal cardiac events [21, 22]. However, in the ORIGIN trial (Outcome Reduction With Initial Glargine Intervention) [23], the 12536 patients randomized to omega-3 PUFA supplement or placebo were at high risk of CVD, with the majority having had a prior CHD event. Supplementation with omega-3 PUFA did not affect the risk of death resulting from cardiovascular causes or the other prespecified secondary cardiovascular outcomes. Another large general-practice cohort of patients with multiple cardiovascular risk factors, daily treatment with n-3 fatty acids (1g daily) did not reduce cardiovascular mortality and morbidity [24]. Similar negative results were seen in acute myocardial infarction [25]. Meanwhile, a simple and safe treatment with n-3 PUFA can provide a small beneficial advantage in terms of mortality and admission to hospital for cardiovascular reasons in patients with heart failure [26].

Risk and Prevention Risk and Prevention Study Collaborative Group reported that patients with prior CVD more than 4 risk factors (more than 65years old, male, hypertension, high cholesterol, current smoker, obesity, family history of CVD) or diabetes mellitus and more than 1CVD but not myocardial infarction showed no benefit from omega-3 PUFA supplementation (850 mg/day EPA+DHA). In AREDS2 randomized clinical trials [27], no reduction in the risk of CVD or secondary CVD outcomes was seen for the DHA+EPA (350 mg DHA+ 650 mg EPA, daily), (primary outcome: hazard ratio [HR], 0.95; 95% CI, 0.78-1.17).

Meanwhile, The JELIS trial (Japan EPA Lipid Intervention Study) reported results by prior CVD status. Although 80% of the patients had no prior clinical CVD, all the patients had a total cholesterol >6.5 mmol/L (>250 mg/dL). Patients were randomly assigned to either statin alone or statin with 1800 mg/d EPA. The primary end was a composite outcome comprising major coronary events, including sudden cardiac death (SCD), fatal and nonfatal myocardial infarction (MI), and other nonfatal events (unstable angina pectoris, angioplasty, stenting, or coronary artery bypass grafting). The composite outcome was reduced among patients treated with the omega-3 PUFA supplement: The relative risk (RR) reduction was 19% (P=0.01) overall, 19% (P=0.048) in those with prior CVD, and 18% (P=0.132) in those without prior CVD (P for interaction=0.95). The risk reduction occurred primarily from a reduction in unstable angina, coronary artery bypass grafting, or percutaneous transluminal coronary angioplasty. There was little evidence of risk reduction in the hard endpoints of nonfatal MI (myocardial infarction) and CHD death.

Overall, there was a lack of consensus on whether current evidence from RCTs suggested no benefit of omega-3 PUFA supplements among patients at high CVD risk except for the result of JELIS.

4.2. PUFA for Secondary Prevention for CVD

The 2 trials in patients with prior CHD, cited in the 2002 statement, suggested a benefit of omega-3 PUFA supplementation [21, 22]. One trial demonstrated a reduction in hard CVD endpoints (CVD death, nonfatal MI, and nonfatal stroke: RR, 0.80; 95% CI, 0.68–0.94) among patients with recent MI (within 3 months), attributable mostly to a reduction in SCD [21]. A second trial found that providing an omega-3 PUFA supplement or fish advice significantly reduced CHD death but not total CHD among patients with recent MI (mean, ≈1.5 months). Of note, only one-third of the patients in the second RCT took an omega-3 PUFA supplement [22].

It was included in the 2002 report, in part because this report focused both on dietary intake and omega-3 PUFA supplementation. Since the 2002 report, 3 additional secondary prevention trials have examined the impact of omega-3 PUFA supplementation on clinical cardiovascular events. Omega (with recent MI within 2 weeks), Alpha Omega trial (history of MI up to 10 years prior) showed no benefit on the clinical cardiovascular event.

On the contrary, A meta-analysis published in 2012 examined the effects of omega-3 PUFA supplementation and dietary intake in 20 RCTs that enrolled patients at high CVD risk or prevalent CHD and patients with an implantable cardioverter-defibrillator (total n=68680) demonstrated a reduction in CHD death (RR, 0.91; 95% CI, 0.85–0.98), possibly as the result of a lower risk of SCD (RR, 0.87; 95% CI, 0.75–1.01) [28]. Taken together, AHA Science advisory concluded that treatment with omega-3 PUFA supplements is reasonable for the secondary prevention of CHD death (Class IIa Recommendation) [29].

However, most prior observational studies have evaluated associations with self-reported PUFA intake, which may be influenced by reporting bias and limited by nutrient database information. The use of objective biomarkers of these fatty acids minimizes reporting bias and allows direct investigation of individual circulating fatty acids. Yet few prior studies have evaluated both dietary and circulating PUFA biomarkers. In a multiethnic cohort of 2837 US adult (whites, Hispanics, African Americans, Chinese Americans), plasma phospholipid PUFAs were measured at baseline (2000–2002) using gas chromatography and dietary PUFAs estimated using a food frequency questionnaire [30]. Incident CVD events (including coronary heart disease and stroke; n=189) were prospectively identified through 2010 during 19,778 person-years of follow-up. In multivariable-adjusted Cox models, circulating n-3 eicosapentaenoic acid and docosahexaenoic acid were inversely associated with incident CVD, with extreme quartile hazard ratios (95% CIs) of 0.49 for eicosapentaenoic acid (0.30 to 0.79; Ptrend=0.01) and 0.39 for docosahexaenoic acid (0.22 to 0.67; Ptrend<0.001). N-3 Docosapentaenoic acid (DPA) was inversely associated with CVD in whites and Chinese, but not in other race/ethnicities (P-interaction=0.01). No significant associations with CVD were observed for circulating n-3 alpha-linolenic acid or n-6 PUFA (linoleic acid, arachidonic acid). Associations with CVD of self-reported dietary PUFA were consistent with those of the PUFA biomarkers. These findings suggest that increased consumption of n-3 PUFA from seafood may prevent CVD development in a multiethnic population.

Moreover, Nishizaki Y et al. [31] compared serum levels of EPA, DHA and arachidonic acid (AA) between patients with and without acute coronary syndrome (ACS). Next, EPA/AA and DHA/AA were estimated to determine the risk factor for ACS. In multivariate logistic regression analysis, the group with the lowest EPA/AA (<0.33) had a great probability of ACS (odds ratio 3.14, 95% confidence interval 1.16 to 8.49) but this was not true for DHA/AA.

4.3. PUFA for Stroke Prevention

Overall, there is no evidence of the benefit of omega-3 PUFA supplementation to reduce the risk of recurrent stroke or other CVDs among patients with a prior stroke.

4.4. PUFA for Heart Failure

In a large RCT [26] among patients with chronic heart failure (New York Heart Association class II-IV), the effects of omega-3 PUFA supplements on the risk of morbidity or mortality were examined. Among patients with clinically diagnosed heart failure (of any type, severity, and cause), omega-3 PUFA supplementation reduced the risk of total mortality (death resulting from any cause) by 9% (RR, 0.91; 95% CI, 0.833–0.998; P=0.041) and the risk of cardiovascular-related hospitalizations or death by 8% (RR, 0.92; 95% CI, 0.849–0.999; P=0.009) after prespecified adjustment for hospitalization for heart failure during the preceding year, previous pacemaker, and aortic stenosis. Notably, 91% of the patients enrolled in the RCT had heart failure with reduced ejection fraction (left ventricular ejection fraction <40%). Based on this large RCT, treatment with omega-3 PUFA supplements is reasonable among patients with heart failure with reduced ejection fraction.

4.5. Biological Ground of Cardio-Protective Function by PUFA

There are many reports to show biological ground causing the action to control a cardiovascular event by PUFA. Omega-3 PUFAs have a broad range of beneficial CV effects including reducing triglycerides, very-low-density lipoprotein (VLDL), inflammatory markers, remnant-like lipoparticle cholesterol (RLP-C), oxidized low-density lipoprotein (ox-LDL), heart rate, blood pressure, and possibly arrhythmia risk.

4.5.1. Reducing Plasma Triglycerides

Lowering of plasma triglycerides by PUFA is well recognized. Reduced hepatic very low-density lipoprotein synthesis contributes to this effect, with implicated mechanisms including reduced fatty acid availability for triglyceride synthesis due to decrease de novo lipogenesis (the process of converting carbohydrates into fat), increased fatty acid beta-oxidation, and reduced delivery of no esterified fatty acids to the liver; reduced hepatic enzyme activity for triglyceride synthesis [32].

4.5.2. Remnant Lipoprotein

Icosapent ethyl (what is used to lower triglycerides) 4 g/day significantly reduced directly measured RLP-C levels 29.8% (p=0.004) in MARINE [33] study and 25.8% (p=0.0001) in ANCHOR study [33] versus placebo, and reduced directly measured RLP-C levels in subgroups with higher versus lower baseline TG levels, in patients receiving statins versus no statins (MARINE), and in patients receiving medium/higher-intensity versus lower-intensity statins (ANCHOR). Strong correlations were found between calculated and directly measured RLP-C for baseline, end-of-treatment, and percent change values in ANCHOR and MARINE (0.73-0.92; $p<0.0001$ for all) [33]. In another study, EPA (1.8 g/day for 3 months) significantly reduced small dense LDL particles ($P < 0.01$) and RLP-triglycerides ($P < 0.05$) compared with baseline in patients with type 2 diabetes mellitus and the metabolic syndrome, which may have been due in part to a reduction in cholesteryl ester transfer protein activity [34].

4.5.3. Effects of EPA on Endothelial Function/Dysfunction

There is substantial evidence that EPA has a beneficial effect on endothelial function. The release of nitric oxide (NO) by vascular endothelial cells serves to regulate vasomotor tone in response to acetylcholine and other vasoactive agonists [35].

EPA attenuated palmitic acid-induced generation of reactive oxygen species, expression of adhesion molecules and cytokines, activation of apoptosis-related proteins, and apoptosis in HUVECs [36]. At elevated levels, the aldose sugar glucose produces non-enzymatic chemical modifications to membrane proteins and phospholipids, leading to advanced glycation endproducts (AGEs) and cell injury. Oxidative stress and AGEs have been implicated in both microvascular and macrovascular complications of diabetes and other metabolic disorders. In membranes enriched with PUFAs, hyperglycemia promotes the formation of free radicals and cholesterol crystalline domains associated with atherosclerosis. The non-enzymatic effects of glucose on cholesterol crystalline domain formation were shown to be enhanced under conditions of high cholesterol. Oxidative damage to PUFAs with glucose is of interest given its role in the propagation of free radicals during vascular injury and insulin resistance. EPA significantly inhibited glucose-induced lipid peroxidation and cholesterol crystalline domain formation in membrane lipid vesicles. These antioxidant effects may be attributed to the intercalation of EPA into the membrane lipid bilayer where it may interfere with the propagation of reactive oxygen species and preserve membrane lipid structural organization [37].

4.5.4. Effects of EPA on Monocytes, Macrophages, and Foam Cells

EPA has been shown to have beneficial effects on each of these cell types. In hyperlipidemic patients with type 2 diabetes mellitus, EPA (1.8 g/day for 6 months) significantly increased circulating levels of adiponectin, a protein that stimulates NO production and suppresses monocyte attachment to endothelial cells [38]. In experimental models, EPA reduced monocyte adhesion whereas the omega-6 AA increased monocyte adhesion to endothelial cells in the presence or absence of an inflammatory stimulus (i.e., tumor necrosis factor-a [TNFa]) [39]. In an in vitro assay under physiologic flow conditions, EPA inhibited lipopolysaccharide (LPS)-induced and TNFa-induced monocyte rolling and adhesion to HUVECs [40]. EPA inhibited LPS-induced intracellular signaling pathways, leading to a reduction in vascular cell adhesion molecule-1 expression [40]. EPA significantly reduced ox-LDL in patients with very high triglycerides and in statin-treated patients with high triglycerides [41]. This might result in better clearance of LDL and contribute to the differential effects observed in patients for EPA and DHA, where EPA results in no change or a decrease in LDL-C while DHA increases LDL-C and HDL-C. Also, EPA and DHA decrease triglycerides [42]. While RLPs cause foam cell formation. As noted above, EPA has been shown to reduce RLPs in several studies [43, 44].

4.5.5. Effects of EPA on Inflammation and Cytokines

EPA may also promote resolution of vascular inflammation by producing resolvins and protectins [45, 46]; resolvins are formed by aspirin-acetylated COX-2 in vascular endothelial cells and 5-LOX in leukocytes whereas protectins are formed by 15-LOX in multiple cell types [47]. Both resolvins and protectins reduce neutrophil recruitment, thereby helping to resolve inflammatory processes and correct the impaired resolution of vascular inflammation seen in atherosclerosis. EPA has been demonstrated to have a favorable impact on markers of inflammation in clinical studies. In the ANCHOR and MARINE trials, EPA (4 g/day for 12 weeks) significantly reduced high-sensitivity C-reactive protein (hs-CRP) by 22% in statin-treated patients with high triglycerides ($P<0.0005$) and by

36% in patients with very high triglycerides (P<0.0012) compared with placebo [41]; similar results were observed in a subanalysis of patients from MARINE and ANCHOR with metabolic syndrome [48]. Dangardt F et al. reported that Twenty-five obese adolescents (14 females, 11 males, age 15.7±1.0 years, BMI 33.8±3.9) were randomized to receive capsules containing either 1.2g/day n-3 or placebo for 3 months. N-3 supplementation also decreased the lymphocyte, monocyte, TNF-α, IL-6 and IL-1β levels. Thus, omega-3 improved vascular function and reduced inflammation in these obese adolescents. In dyslipidemic obese adults, EPA (1.8 g/day for 3 months) significantly increased serum levels of the anti-inflammatory cytokine IL-10 (P < 0.01) and IL-10 expression by peripheral blood monocytes (P < 0.05) compared with untreated controls [49].

4.5.6. PUFA for Plaque Composition

Nishio et al. reported that the EPA + statin group had higher EPA/AA ratios and lower PTX3 levels (a marker of arterial inflammation) than the statin group. EPA/AA was inversely and significantly correlated with the %Plaque burden (R=−0.261, P=0.048) [52]. Moreover, a low EPA/AA ratio was identified as an independent risk factor for ACS [50]. In the virtual histology intravascular ultrasound analysis, the percentage of dense calcium in the plaque was positively correlated with EPA, EPA/AA and DHA/AA [50]. EPA/AA was inversely and significantly correlated with the %Plaque burden (R=−0.261, P=0.048) [50].

Fibrous cap thickening may help stabilize sites of atherosclerotic plaques and thereby prevent plaque rupture. EPA (1.8 g/day for 8 months) significantly increased fibrous cap thickness in ACS patients compared with baseline and compared with untreated controls as measured by optical coherence tomography (both P < 0.001) [51]. Similarly, OCT analysis by Nishio et al. showed that the EPA + statin group had a greater increase in fibrous-cap thickness [52].

5. PUFA and Inflammatory Bowel Disease (IBD)

It is currently well accepted that on the basis of IBD pathogenesis, [53, 54] there is an imbalance between pro- and anti-inflammatory signals [55]. This suggests that defects in the proper release of pro-resolving molecules during the acute phase of inflammation may characterize IBD onset. For decades, the resolution of inflammation has been considered a passive event, in which pro-inflammatory signals simply dilute over time. This concept was overturned when Serhan and colleagues discovered for the first time that a specific class of lipids, known as eicosanoids and docosanoids, promotes, and orchestrates the resolution process [56, 57]. This discovery gave rise to a new field of research studying the mechanisms and the factors involved in the resolution phase of the inflammatory response, which is finely and temporally regulated by specialized pro-resolving mediators, named lipoxins (LXs), resolvins (Rv), protectins, and maresins. These resolving bioactive lipids are synthesized from n-6 and n-3 PUFAs and have been demonstrated to exert potent immune-resolving effects [58]. PUFAs have been shown to modulate other biological activities of intestinal epithelial cells. It is known that pro-resolving lipid mediators (LMs) exert their functions by binding with cell surface receptors, the majority of which belongs to G protein-coupled receptors (GPRs) [59]. Among these receptors, GPR120 has been found to be the most abundantly expressed in the gut, particularly on epithelial cells and macrophages [59]. A study from Mobraten and colleagues shows that DHA, EPA, or AA are able to trigger GPR120 in Caco-2 cells, initiating multiple and independent signaling processes with different kinetics and intensity; these are (i) the activation of MAP kinases, (ii) the inhibition of IL-1β induced NF-κB activation, and (iii) the cytosolic accumulation of Ca^{2+} [60]. Another group shows differential effects of activation of GPR120 by DHA in human intestinal Caco-2 and murine STC-1 cells, two different cell lines representing the mammalian intestinal epithelial layer. In this study, GPR120 stimulation by n-3 PUFAs increased β-arrestin2 interaction with TAB 1 and attenuated TNFα-induced inflammatory effects by the association of TAB 1 with TAK1, which resulted in reduced activation of NF-kB [61]. Anti-inflammatory effects exerted by PUFAs through GPR120 were confirmed *in vivo* by Zhao et al., who demonstrated that triggering of GPR120 by DHA treatment ameliorate the experimental colitis in IL-10 deficient mice [59]. Interestingly, transcription of *GPR120* in intestinal epithelial cells was found tremendously increased by bacteria belonging to the *Firmicutes, Bacteroides and Proteobacteria phyla* [62], all classified as microorganisms harboring anti-inflammatory properties. Matsuoka et al. [63] suggests that reduced expression levels of GPR120 may be one of the causes underlining IBD pathogenesis and that targeting this receptor may represent a new therapeutic strategy in IBD; however, to date there are no studies that deeply characterize and quantify GPR120 in the inflamed mucosa of IBD patients and further studies to elucidate this aspect are needed.

John et al. found that the intake of dietary EPA and DHA was conversely correlated with the risk of developing incident UC (Ulcerative Colitis) [64]. Similarly, in a cohort of patients with CD (Crohn's Disease), the dietary DHA intake was conversely correlated with the development of incident CD, with statistical significance [65]. Moreover, clinical trial for CD and UC revealed the beneficial effects of n-3 enriched diet [66, 67, 68, 69]. However, negative meta-analysis reports [70, 71] described that there are insufficient data to recommend the use of omega 3 fatty acids for maintenance of remission in CD and UC. There was still controversy about the clinical efficacy of PUFA for IBD.

6. PUFA and Alzheimer Disease

The central nervous system (CNS) is highly enriched in n-3 LC-PUFAs, of which DHA is the most important, participating in many crucial processes within the mammalian cells, as well as being the most abundant component of the neuron membrane, having important structural and functional roles. Membrane status affects the transfer of neuronal information, speed of signal transduction, and interaction with proteins, indicating the importance of this FA in order to gain appropriate functioning [72], DHA is also the principal n-3 FA in cortical gray matter, representing ~15% of total FAs in the adult human prefrontal cortex (PFC). Moreover, other FAs like EPA, and less importantly DPA, are also present in the CNS, constituting around 1% of total brain FA composition [73].

Alzheimer's disease (AD) is a neurodegenerative disorder that produces severe cognitive impairment as it progresses. This pathology is the main neurological cause of dementia suffered by 36 million people worldwide, elderly adults in most cases (World Alzheimer Report 2011) [74]. The molecular phenotype is characterized by the accumulation of amyloid β peptide (Aβ) and aggregates composed of the hyperphosphorylated microtubule-associated tau protein [75]. Regarding AD, statistical differences in the amount of DHA in the liver of control subjects (324.83 ± 122.89 μg/ml) and those of patients with Alzheimer's disease (204.64 ± 74.62 μg/ml) have been reported. These data correlated with both a lower amount of DHA in the brain and a lower expression level of two genes involved in DHA biosynthesis in the liver, fds2 and helo1 [76].

On the other hand, EPA, which commonly displays 250–300 times lower level in the brain compared to DHA, appears preferably esterified to phosphatidylinositol (PI) [77] and plays also an important role in the AD. It was shown that higher plasma levels of EPA, but not DHA, were connected with lower gray matter atrophy, suggesting that EPA might play an important role in cognitive decline. Higher EPA levels were also found in the right amygdala and may be associated with depressive symptoms [78]. Grimm M et al. reported that DHA and EPA mediate degradation of Aβ by stimulating and boosting the gene expression of insulin-degrading enzyme (ide), resulting in a higher amount of IDE, the major Aβ-degrading enzyme secreted into the extracellular space of neuronal and microglial cells [79]. Furthermore, it has been reported that fat-1 transgenic mice, which express more n-3 PUFAs (total n-3 PUFAs in WT: 17.51, in fat-1 mice: 23.45), are more resistant to intracerebroventricular injection of Aβ, which induces neuronal cell loss and impairment of hippocampus-dependent memory [80]. Similar to the results found in fat-1 mice, the clearance of interstitial Aβ was much faster in those mice treated with fish oil, compared to untreated mice. Such a protective effect has been suggested to occur through an activation of the AQP4-mediated glymphatic system to promote interstitial Aβ clearance [81]. DHA, as a fundamental component of neuronal membranes, has been widely studied as an active compound to treat Alzheimer effects [82]. It has been shown that the novel hydroxyl-derivative of DHA (OHDHA) has a strong therapeutic potential to treat AD. It has been demonstrated that administration of OHDHA increases the concentration of PUFA's containing phospholipids (e.g., DHA or EPA-containing phospholipids). The increase in n-3 PUFAs in the transgenic mouse model of the AD was accompanied by a parallel reduction in amyloid-β accumulation, which is strongly correlated with synaptic degeneration, and with memory failure and learning deficiencies [74].

7. PUFA and Parkinson Disease

The relationship between PUFA and Parkinson disease (PD) has not been elucidated. One study investigated the effect of the neurotoxin 1-methyl-4-phenyl-1,2,3,6-tetrahydropyridine (MPTP) when given to mice (from 2 to 12 months of age) fed a diet high in n-3 PUFAs. It was observed that the rich n-3 PUFA diet protected against the MPTP-induced decrease in dopamine, and its metabolite dihydroxyphenylacetic acid in the striatum. These data suggest that a high n-3 PUFAs dietary intake might exert neuroprotective actions in an animal model of PD [83]. Within LC-PUFAs, DHA is the most important one for the brain; recently, in an experimental mouse model of PD the effects of the neurotoxin MPTP administrated together with DHA were evaluated. The results indicated that DHA has protective effects on dopaminergic neurons in this MPTP-induced experimental model [84]. Transgenic animal models have also been used to test the protective effect of high levels of n-3 PUFAs. Bousquet and collaborators published that the expression of the fat-1 transgene (which encodes the enzyme n-3 PUFA desaturase that catalyzes the n-6 to n-3 PUFA conversion) in mice increased the cortical n-3: n-6 PUFA ratio

(+28%) compared to control. This modification is less effective than a dietary supplementation since a 10-month exposure to an n-3 PUFA-enriched diet increased the n-3:n-6 PUFA ratio by as much as 92%. Regarding neuroprotection in animal models of PD using MPTP, the data suggest that dietary intake of a preformed DHA supplement is more effective than the expression of the fat-1 transgene [85]. The preventing effect of DHA on dopaminergic lesion was also studied by determining the levels of tyrosine hydroxylase (TH) in MPTP and MPTP + DHA treated cells. There was a 65% decline in the number of TH-immunopositive neurons in the substantia nigra one week after MPTP administration and the number of TH-positive neurons significantly increased in the MPTP + DHA group compared to the MPTP control group [86].

8. PUFA and Amyotrophic Lateral Sclerosis (ALS)

A recent study in a mouse model of ALS (G93A mutated human superoxide dismutase) has identified a relationship between neuronal loss and DHA levels. In this mouse model a significant decrease in DHA-containing phosphatidylcholines (PCs), in the terminal stage of ALS was shown [87]. A dietary intake of n-3 PUFA is important in ALS, as well as in Alzheimer or Parkinson. In a large analysis of a total of 995 ALS patients fed with an n-3 PUFA enriched diet, this intake was associated with a risk reduction of 34% for ALS. Further analysis of the individual n-3 PUFAs established that intakes of ALA and marine n-3 PUFAs were each associated with a lower ALS risk [88].

Microglia is considered the "resident macrophages" of the brain. When in their resting state, microglia performs routine maintenance and immune surveillance. Once activated, either by injury or an immune stimulus, microglia secretes a variety of pro-inflammatory molecules, such as Nitric Oxide, superoxide, and inflammatory cytokines. Up-regulation of pro-inflammatory molecules is transient and does not cause neurodegeneration. However, if up-regulation lasts for an extended period, neurodegeneration ensues. Many neurodegenerative diseases are characterized by chronic inflammation due to microglial activation. DHA increased total glutathione levels in microglial cells and enhanced their anti-oxidative capacity. It reduced production of the pro-inflammatory cytokines TNF-α and IL-6 induced through TLR-3 and TLR-4 activation. Furthermore, it reduced production of Nitric Oxide [89].

CONCLUSION

Many biological findings and clinical evidence of implication between PUFA and human health have been elucidated one after another in various regions. Consistently, n-3 PUFA has the beneficial effect for many tissues in the human bodies (i.e., brain, cardiovascular system, and bowels). We should intake more n-3 PUFA to keep a better status of organs from the supplement, oil, plant and/or fish. Moreover, further studies about the influence for human health by PUFA should be exerted to clarify. Those accumulations will bring us the resolutions for the intractable disease in the future.

REFERENCES

[1] Stillwell W, Wassall SR. Docosahexaenoic acid: membrane properties of a unique fatty acid. *Chemistry and physics of lipids.* 2003;126(1):1-27.

[2] Lauritzen I, Blondeau N, Heurteaux C, Widmann C, Romey G, Lazdunski M. Polyunsaturated fatty acids are potent neuroprotectors. *The EMBO journal.* 2000;19(8):1784-93.

[3] Moreno C, Macias A, Prieto A, de la Cruz A, Gonzalez T, Valenzuela C. Effects of n-3 Polyunsaturated Fatty Acids on Cardiac Ion Channels. *Frontiers in physiology.* 2012;3:245.

[4] Jacques C, Levy E, Muckle G, Jacobson SW, Bastien C, Dewailly E, et al. Long-term effects of prenatal omega-3 fatty acid intake on

visual function in school-age children. *J Pediatr*. 2011;158(1):83-90, e1.

[5] Johansson I, Monsen VT, Pettersen K, Mildenberger J, Misund K, Kaarniranta K, et al. The marine n-3 PUFA DHA evokes cytoprotection against oxidative stress and protein misfolding by inducing autophagy and NFE2L2 in human retinal pigment epithelial cells. *Autophagy*. 2015;11(9):1636-51.

[6] Merle BM, Benlian P, Puche N, Bassols A, Delcourt C, Souied EH. Circulating omega-3 Fatty acids and neovascular age-related macular degeneration. *Invest Ophthalmol Vis Sci*. 2014;55(3):2010-9.

[7] Sasaki M, Kawasaki R, Rogers S, Man RE, Itakura K, Xie J, et al. The Associations of Dietary Intake of Polyunsaturated Fatty Acids with Diabetic Retinopathy in Well-Controlled Diabetes. *Invest Ophthalmol Vis Sci*. 2015;56(12):7473-9.

[8] Albanese E, Dangour AD, Uauy R, Acosta D, Guerra M, Guerra SS, et al. Dietary fish and meat intake and dementia in Latin America, China, and India: a 10/66 Dementia Research Group population-based study. *Am J Clin Nutr*. 2009;90(2):392-400.

[9] Barberger-Gateau P, Raffaitin C, Letenneur L, Berr C, Tzourio C, Dartigues JF, et al. Dietary patterns and risk of dementia: the Three-City cohort study. *Neurology*. 2007;69(20):1921-30.

[10] Kalmijn S, Launer LJ, Ott A, Witteman JC, Hofman A, Breteler MM. Dietary fat intake and the risk of incident dementia in the Rotterdam Study. *Ann Neurol*. 1997;42(5):776-82.

[11] Morris MC, Evans DA, Bienias JL, Tangney CC, Bennett DA, Wilson RS, et al. Consumption of fish and n-3 fatty acids and risk of incident Alzheimer disease. *Arch Neurol*. 2003;60(7):940-6.

[12] Devore EE, Grodstein F, van Rooij FJ, Hofman A, Rosner B, Stampfer MJ, et al. Dietary intake of fish and omega-3 fatty acids in relation to long-term dementia risk. *Am J Clin Nutr*. 2009;90(1):170-6.

[13] Lin PY, Chiu CC, Huang SY, Su KP. A meta-analytic review of polyunsaturated fatty acid compositions in dementia. *J Clin Psychiatry*. 2012;73(9):1245-54.

[14] Mazereeuw G, Lanctot KL, Chau SA, Swardfager W, Herrmann N. Effects of omega-3 fatty acids on cognitive performance: a meta-analysis. *Neurobiol Aging.* 2012;33(7):1482.e17-29.

[15] Sydenham E, Dangour AD, Lim WS. Omega 3 fatty acid for the prevention of cognitive decline and dementia. *Cochrane Database Syst Rev.* 2012(6):Cd005379.

[16] Wu S, Ding Y, Wu F, Li R, Hou J, Mao P. Omega-3 fatty acids intake and risks of dementia and Alzheimer's disease: a meta-analysis. *Neurosci Biobehav Rev.* 2015;48:1-9.

[17] Solfrizzi V, Custodero C, Lozupone M, Imbimbo BP, Valiani V, Agosti P, et al. Relationships of Dietary Patterns, Foods, and Micro- and Macronutrients with Alzheimer's Disease and Late-Life Cognitive Disorders: A Systematic Review. *J Alzheimers Dis.* 2017;59(3):815-49.

[18] Zhang Y, Chen J, Qiu J, Li Y, Wang J, Jiao J. Intakes of fish and polyunsaturated fatty acids and mild-to-severe cognitive impairment risks: a dose-response meta-analysis of 21 cohort studies. *Am J Clin Nutr.* 2016;103(2):330-40.

[19] Perez MA, Penaloza-Sancho V, Ahumada J, Fuenzalida M, Dagnino-Subiabre A. n-3 Polyunsaturated fatty acid supplementation restored impaired memory and GABAergic synaptic efficacy in the hippocampus of stressed rats. *Nutr Neurosci.* 2017:1-14.

[20] Kris-Etherton PM, Harris WS, Appel LJ. Fish consumption, fish oil, omega-3 fatty acids, and cardiovascular disease. *Circulation.* 2002;106(21):2747-57.

[21] Dietary supplementation with n-3 polyunsaturated fatty acids and vitamin E after myocardial infarction: results of the GISSI-Prevenzione trial. Gruppo Italiano per lo Studio della Sopravvivenza nell'Infarto miocardico. *Lancet.* 1999;354(9177):447-55.

[22] Burr ML, Fehily AM, Gilbert JF, Rogers S, Holliday RM, Sweetnam PM, et al. Effects of changes in fat, fish, and fibre intakes on death and myocardial reinfarction: diet and reinfarction trial (DART). *Lancet.* 1989;2(8666):757-61.

[23] Bosch J, Gerstein HC, Dagenais GR, Diaz R, Dyal L, Jung H, et al. n-3 fatty acids and cardiovascular outcomes in patients with dysglycemia. *N Engl J Med.* 2012;367(4):309-18.

[24] Roncaglioni MC, Tombesi M, Avanzini F, Barlera S, Caimi V, Longoni P, et al. n-3 fatty acids in patients with multiple cardiovascular risk factors. *N Engl J Med*. 2013;368(19):1800-8.

[25] Rauch B, Schiele R, Schneider S, Diller F, Victor N, Gohlke H, et al. OMEGA, a randomized, placebo-controlled trial to test the effect of highly purified omega-3 fatty acids on top of modern guideline-adjusted therapy after myocardial infarction. *Circulation.* 2010;122(21):2152-9.

[26] Tavazzi L, Maggioni AP, Marchioli R, Barlera S, Franzosi MG, Latini R, et al. Effect of n-3 polyunsaturated fatty acids in patients with chronic heart failure (the GISSI-HF trial): a randomised, double-blind, placebo-controlled trial. *Lancet.* 2008;372(9645):1223-30.

[27] Bonds DE, Harrington M, Worrall BB, Bertoni AG, Eaton CB, Hsia J, et al. Effect of long-chain omega-3 fatty acids and lutein + zeaxanthin supplements on cardiovascular outcomes: results of the Age-Related Eye Disease Study 2 (AREDS2) randomized clinical trial. *JAMA Intern Med.* 2014;174(5):763-71.

[28] Rizos EC, Ntzani EE, Bika E, Kostapanos MS, Elisaf MS. Association between omega-3 fatty acid supplementation and risk of major cardiovascular disease events: a systematic review and meta-analysis. *Jama.* 2012;308(10):1024-33.

[29] Siscovick DS, Barringer TA, Fretts AM, Wu JH, Lichtenstein AH, Costello RB, et al. Omega-3 Polyunsaturated Fatty Acid (Fish Oil) Supplementation and the Prevention of Clinical Cardiovascular Disease: A Science Advisory From the American Heart Association. *Circulation.* 2017;135(15):e867-e84.

[30] de Oliveira Otto MC, Wu JH, Baylin A, Vaidya D, Rich SS, Tsai MY, et al. Circulating and dietary omega-3 and omega-6 polyunsaturated fatty acids and incidence of CVD in the Multi-Ethnic Study of Atherosclerosis. *J Am Heart Assoc.* 2013;2(6):e000506.

[31] Nishizaki Y, Shimada K, Tani S, Ogawa T, Ando J, Takahashi M, et al. Significance of imbalance in the ratio of serum n-3 to n-6 polyunsaturated fatty acids in patients with acute coronary syndrome. *The American journal of cardiology*. 2014;113(3):441-5.

[32] Agrawal N, Freitas Corradi P, Gumaste N, Goldberg IJ. Triglyceride Treatment in the Age of Cholesterol Reduction. *Progress in cardiovascular diseases*. 2016;59(2):107-18.

[33] Ballantyne CM, Bays HE, Philip S, Doyle RT, Jr., Braeckman RA, Stirtan WG, et al. Icosapent ethyl (eicosapentaenoic acid ethyl ester): Effects on remnant-like particle cholesterol from the MARINE and ANCHOR studies. *Atherosclerosis*. 2016;253:81-7.

[34] Satoh N, Shimatsu A, Kotani K, Sakane N, Yamada K, Suganami T, et al. Purified eicosapentaenoic acid reduces small dense LDL, remnant lipoprotein particles, and C-reactive protein in metabolic syndrome. *Diabetes care*. 2007;30(1):144-6.

[35] Tousoulis D, Kampoli AM, Tentolouris C, Papageorgiou N, Stefanadis C. The role of nitric oxide on endothelial function. *Current vascular pharmacology*. 2012;10(1):4-18.

[36] Lee CH, Lee SD, Ou HC, Lai SC, Cheng YJ. Eicosapentaenoic acid protects against palmitic acid-induced endothelial dysfunction via activation of the AMPK/eNOS pathway. *International journal of molecular sciences*. 2014;15(6):10334-49.

[37] Mason RP, Jacob RF. Eicosapentaenoic acid inhibits glucose-induced membrane cholesterol crystalline domain formation through a potent antioxidant mechanism. *Biochimica et biophysica acta*. 2015;1848(2):502-9.

[38] Nomura S, Shouzu A, Omoto S, Inami N, Ueba T, Urase F, et al. Effects of eicosapentaenoic acid on endothelial cell-derived microparticles, angiopoietins and adiponectin in patients with type 2 diabetes. *Journal of atherosclerosis and thrombosis*. 2009;16(2):83-90.

[39] Grenon SM, Aguado-Zuniga J, Hatton JP, Owens CD, Conte MS, Hughes-Fulford M. Effects of fatty acids on endothelial cells:

inflammation and monocyte adhesion. *The Journal of surgical research*. 2012;177(1):e35-43.

[40] Yamada H, Yoshida M, Nakano Y, Suganami T, Satoh N, Mita T, et al. In vivo and in vitro inhibition of monocyte adhesion to endothelial cells and endothelial adhesion molecules by eicosapentaenoic acid. *Arteriosclerosis, thrombosis, and vascular biology*. 2008;28(12):2173-9.

[41] Bays HE, Ballantyne CM, Braeckman RA, Stirtan WG, Soni PN. Icosapent ethyl, a pure ethyl ester of eicosapentaenoic acid: effects on circulating markers of inflammation from the MARINE and ANCHOR studies. *Am J Cardiovasc Drugs*. 2013;13(1):37-46.

[42] Wei MY, Jacobson TA. Effects of eicosapentaenoic acid versus docosahexaenoic acid on serum lipids: a systematic review and meta-analysis. *Current atherosclerosis reports*. 2011;13(6):474-83.

[43] Nakamura N, Hamazaki T, Kobayashi M, Ohta M, Okuda K. Effects of eicosapentaenoic acids on remnant-like particles, cholesterol concentrations and plasma fatty acid composition in patients with diabetes mellitus. *In vivo* (Athens, Greece). 1998;12(3):311-4.

[44] Hamazaki K, Itomura M, Huan M, Nishizawa H, Watanabe S, Hamazaki T, et al. n-3 long-chain FA decrease serum levels of TG and remnant-like particle-cholesterol in humans. *Lipids*. 2003;38(4):353-8.

[45] Spite M, Claria J, Serhan CN. Resolvins, specialized proresolving lipid mediators, and their potential roles in metabolic diseases. *Cell metabolism*. 2014;19(1):21-36.

[46] Hasturk H, Abdallah R, Kantarci A, Nguyen D, Giordano N, Hamilton J, et al. Resolvin E1 (RvE1) Attenuates Atherosclerotic Plaque Formation in Diet and Inflammation-Induced Atherogenesis. *Arteriosclerosis, thrombosis, and vascular biology*. 2015;35(5):1123-33.

[47] Adkins Y, Kelley DS. Mechanisms underlying the cardioprotective effects of omega-3 polyunsaturated fatty acids. *The Journal of nutritional biochemistry*. 2010;21(9):781-92.

[48] Bays HE, Ballantyne CM, Braeckman RA, Stirtan WG, Doyle RT, Jr., Philip S, et al. Icosapent Ethyl (Eicosapentaenoic Acid Ethyl Ester): Effects Upon High-Sensitivity C-Reactive Protein and Lipid Parameters in Patients with Metabolic Syndrome. *Metabolic syndrome and related disorders.* 2015;13(6):239-47.

[49] Satoh-Asahara N, Shimatsu A, Sasaki Y, Nakaoka H, Himeno A, Tochiya M, et al. Highly purified eicosapentaenoic acid increases interleukin-10 levels of peripheral blood monocytes in obese patients with dyslipidemia. *Diabetes care.* 2012;35(12):2631-9.

[50] Ohwada T, Yokokawa T, Kanno Y, Hotsuki Y, Sakamoto T, Watanabe K, et al. Vascular composition data supporting the role of N-3 polyunsaturated fatty acids in the prevention of cardiovascular disease events. *Data in brief.* 2016;7:1237-47.

[51] Yamano T, Kubo T, Shiono Y, Shimamura K, Orii M, Tanimoto T, et al. Impact of eicosapentaenoic acid treatment on the fibrous cap thickness in patients with coronary atherosclerotic plaque: an optical coherence tomography study. *Journal of atherosclerosis and thrombosis.* 2015;22(1):52-61.

[52] Nishio R, Shinke T, Otake H, Nakagawa M, Nagoshi R, Inoue T, et al. Stabilizing effect of combined eicosapentaenoic acid and statin therapy on coronary thin-cap fibroatheroma. *Atherosclerosis.* 2014;234(1):114-9.

[53] Das UN. Inflammatory bowel disease as a disorder of an imbalance between pro- and anti-inflammatory molecules and deficiency of resolution bioactive lipids. *Lipids in health and disease.* 2016;15:11.

[54] Ungaro F, Tacconi C, Massimino L, Corsetto PA, Correale C, Fonteyne P, et al. MFSD2A Promotes Endothelial Generation of Inflammation-Resolving Lipid Mediators and Reduces Colitis in Mice. *Gastroenterology.* 2017;153(5):1363-77.e6.

[55] Neurath MF. Cytokines in inflammatory bowel disease. *Nature reviews Immunology.* 2014;14(5):329-42.

[56] Serhan CN. Novel eicosanoid and docosanoid mediators: resolvins, docosatrienes, and neuroprotectins. *Current opinion in clinical nutrition and metabolic care.* 2005;8(2):115-21.

[57] Levy BD, Clish CB, Schmidt B, Gronert K, Serhan CN. Lipid mediator class switching during acute inflammation: signals in resolution. *Nature immunology*. 2001;2(7):612-9.

[58] Serhan CN, Chiang N, Dalli J. The resolution code of acute inflammation: Novel pro-resolving lipid mediators in resolution. *Seminars in immunology*. 2015;27(3):200-15.

[59] Zhao J, Wang H, Shi P, Wang W, Sun Y. GPR120, a potential therapeutic target for experimental colitis in IL-10 deficient mice. *Oncotarget*. 2017;8(5):8397-405.

[60] Mobraten K, Haug TM, Kleiveland CR, Lea T. Omega-3 and omega-6 PUFAs induce the same GPR120-mediated signalling events, but with different kinetics and intensity in Caco-2 cells. *Lipids in health and disease*. 2013;12:101.

[61] Anbazhagan AN, Priyamvada S, Gujral T, Bhattacharyya S, Alrefai WA, Dudeja PK, et al. A novel anti-inflammatory role of GPR120 in intestinal epithelial cells. *American journal of physiology Cell physiology*. 2016;310(7):C612-21.

[62] Fredborg M, Theil PK, Jensen BB, Purup S. G protein-coupled receptor120 (GPR120) transcription in intestinal epithelial cells is significantly affected by bacteria belonging to the Bacteroides, Proteobacteria, and Firmicutes phyla. *Journal of animal science*. 2012;90 Suppl 4:10-2.

[63] Matsuoka K, Kanai T. The gut microbiota and inflammatory bowel disease. *Seminars in immunopathology*. 2015;37(1):47-55.

[64] John S, Luben R, Shrestha SS, Welch A, Khaw KT, Hart AR. Dietary n-3 polyunsaturated fatty acids and the aetiology of ulcerative colitis: a UK prospective cohort study. *European journal of gastroenterology & hepatology*. 2010;22(5):602-6.

[65] Chan SS, Luben R, Olsen A, Tjonneland A, Kaaks R, Lindgren S, et al. Association between high dietary intake of the n-3 polyunsaturated fatty acid docosahexaenoic acid and reduced risk of Crohn's disease. *Alimentary pharmacology & therapeutics*. 2014;39(8):834-42.

[66] Barbosa DS, Cecchini R, El Kadri MZ, Rodriguez MA, Burini RC, Dichi I. Decreased oxidative stress in patients with ulcerative colitis supplemented with fish oil omega-3 fatty acids. *Nutrition* (Burbank, Los Angeles County, Calif). 2003;19(10):837-42.

[67] Stenson WF, Cort D, Rodgers J, Burakoff R, DeSchryver-Kecskemeti K, Gramlich TL, et al. Dietary supplementation with fish oil in ulcerative colitis. *Annals of internal medicine.* 1992;116(8):609-14.

[68] Romano C, Cucchiara S, Barabino A, Annese V, Sferlazzas C. Usefulness of omega-3 fatty acid supplementation in addition to mesalazine in maintaining remission in pediatric Crohn's disease: a double-blind, randomized, placebo-controlled study. *World journal of gastroenterology.* 2005;11(45):7118-21.

[69] Belluzzi A, Brignola C, Campieri M, Pera A, Boschi S, Miglioli M. Effect of an enteric-coated fish-oil preparation on relapses in Crohn's disease. *N Engl J Med.* 1996;334(24):1557-60.

[70] Barbalho SM, Goulart Rde A, Quesada K, Bechara MD, de Carvalho Ade C. Inflammatory bowel disease: can omega-3 fatty acids really help? *Annals of gastroenterology.* 2016;29(1):37-43.

[71] Turner D, Shah PS, Steinhart AH, Zlotkin S, Griffiths AM. Maintenance of remission in inflammatory bowel disease using omega-3 fatty acids (fish oil): a systematic review and meta-analyses. *Inflammatory bowel diseases.* 2011;17(1):336-45.

[72] Cunnane SC, Plourde M, Pifferi F, Begin M, Feart C, Barberger-Gateau P. Fish, docosahexaenoic acid and Alzheimer's disease. *Prog Lipid Res.* 2009;48(5):239-56.

[73] McNamara RK. DHA deficiency and prefrontal cortex neuropathology in recurrent affective disorders. *The Journal of nutrition.* 2010;140(4):864-8.

[74] Torres M, Price SL, Fiol-Deroque MA, Marcilla-Etxenike A, Ahyayauch H, Barcelo-Coblijn G, et al. Membrane lipid modifications and therapeutic effects mediated by hydroxydocosahexaenoic acid on Alzheimer's disease. *Biochimica et biophysica acta.* 2014;1838(6):1680-92.

[75] Blennow K, de Leon MJ, Zetterberg H. Alzheimer's disease. *Lancet* (London, England). 2006;368(9533):387-403.

[76] Astarita G, Jung KM, Berchtold NC, Nguyen VQ, Gillen DL, Head E, et al. Deficient liver biosynthesis of docosahexaenoic acid correlates with cognitive impairment in Alzheimer's disease. *PloS one.* 2010;5(9):e12538.

[77] Dyall SC. Long-chain omega-3 fatty acids and the brain: a review of the independent and shared effects of EPA, DPA and DHA. *Frontiers in aging neuroscience.* 2015;7:52.

[78] Samieri C, Maillard P, Crivello F, Proust-Lima C, Peuchant E, Helmer C, et al. Plasma long-chain omega-3 fatty acids and atrophy of the medial temporal lobe. *Neurology.* 2012;79(7):642-50.

[79] Grimm MO, Mett J, Stahlmann CP, Haupenthal VJ, Blumel T, Stotzel H, et al. Eicosapentaenoic acid and docosahexaenoic acid increase the degradation of amyloid-beta by affecting insulin-degrading enzyme. *Biochemistry and cell biology = Biochimie et biologie cellulaire.* 2016;94(6):534-42.

[80] Ledo JH, Azevedo EP, Clarke JR, Ribeiro FC, Figueiredo CP, Foguel D, et al. Amyloid-beta oligomers link depressive-like behavior and cognitive deficits in mice. *Molecular psychiatry.* 2013;18(10):1053-4.

[81] Ren H, Luo C, Feng Y, Yao X, Shi Z, Liang F, et al. Omega-3 polyunsaturated fatty acids promote amyloid-beta clearance from the brain through mediating the function of the glymphatic system. *FASEB Journal:* official publication of the Federation of American Societies for Experimental Biology. 2017;31(1):282-93.

[82] Oster T, Pillot T. Docosahexaenoic acid and synaptic protection in Alzheimer's disease mice. *Biochimica et biophysica acta.* 2010;1801(8):791-8.

[83] Bousquet M, Saint-Pierre M, Julien C, Salem N, Jr., Cicchetti F, Calon F. Beneficial effects of dietary omega-3 polyunsaturated fatty acid on toxin-induced neuronal degeneration in an animal model of Parkinson's disease. *FASEB Journal:* official publication of the

Federation of American Societies for Experimental Biology. 2008;22(4):1213-25.

[84] Ozkan A, Parlak H, Tanriover G, Dilmac S, Ulker SN, Birsen I, et al. The protective mechanism of docosahexaenoic acid in mouse model of Parkinson: *The role of hemeoxygenase. Neurochemistry international.* 2016.

[85] Bousquet M, Gue K, Emond V, Julien P, Kang JX, Cicchetti F, et al. Transgenic conversion of omega-6 into omega-3 fatty acids in a mouse model of Parkinson's disease. *Journal of lipid research.* 2011;52(2):263-71.

[86] Ozsoy O, Seval-Celik Y, Hacioglu G, Yargicoglu P, Demir R, Agar A, et al. The influence and the mechanism of docosahexaenoic acid on a mouse model of Parkinson's disease. *Neurochemistry International.* 2011;59(5):664-70.

[87] Arima H, Omura T, Hayasaka T, Masaki N, Hanada M, Xu D, et al. Reductions of docosahexaenoic acid-containing phosphatidylcholine levels in the anterior horn of an ALS mouse model. *Neuroscience.* 2015;297:127-36.

[88] Fitzgerald KC, O'Reilly EJ, Falcone GJ, McCullough ML, Park Y, Kolonel LN, et al. Dietary omega-3 polyunsaturated fatty acid intake and risk for amyotrophic lateral sclerosis. *JAMA neurology.* 2014;71(9):1102-10.

[89] Pettit LK, Varsanyi C, Tadros J, Vassiliou E. Modulating the inflammatory properties of activated microglia with Docosahexaenoic acid and Aspirin. *Lipids in health and disease.* 2013;12:16.

In: Polyunsaturated Fatty Acids (PUFAs) ISBN: 978-1-53613-572-5
Editor: Angel Catala

Chapter 3

POLYUNSATURATED FATTY ACIDS IN ARTIFICIAL MEMBRANES MODELED BY THE LANGMUIR MONOLAYER TECHNIQUE

***J. L. Fidalgo Rodríguez*[1,*]*, P. Dynarowicz-Latka*[2]**
***and J. Miñones Conde*[1]**

[1]Department of Physical Chemistry, Faculty of Pharmacy, University of Santiago de Compostela, Spain
[2]Faculty of Chemistry, Jagiellonian University, Kraków, Poland

ABSTRACT

The Langmuir monolayer technique is one of the methods used to model biomembranes and serves as a platform to study the interactions of membrane components with other biomolecules. This method, briefly reviewed in this paper, together with surface visualization method - Brewster angle microscopy (BAM) - has been applied to characterize six unsaturated fatty acids (UFAs), belonging to the omega-9 (oleic acid), omega-3 (α-linolenic and stearidonic acids) and omega-6 (linoleic, γ-linolenic and eicosadienoic acids) in Langmuir monolayers. The

* Corresponding author email: jl_fidalgo@hotmail.com.

influence of such factors as subphase temperature, length of the aliphatic chain, degree of unsaturation and position of the double bonds on the conformation of molecules at the air/water interface and monolayer thickness has been discussed. This paper summarizes the results of interactions between unsaturated fatty acids and major membrane lipids, such as cholesterol and phospholipids. Strong interactions, leading to the formation of stable complexes of unsaturated fatty acids and cholesterol, in which free cholesterol is bound, have been found. In this way, UFAs may impact in lowering cholesterol level. Studies of interactions between unsaturated fatty acids and cholesterol/phospholipid mixed monolayers, mimicking cellular membranes, allowed to conclude that UFAs fluidize model membranes and - present in small amounts - increase their stability.

Keywords: polyunsaturated fatty acids, Langmuir monolayers, interactions, model lipid membranes

ABBREVIATIONS

A,	molecular area;
A/W,	air/water interface;
α-LA,	α-linolenic acid;
BAM,	Brewster angle microscope;
C_s^{-1},	compressibility moduli;
EA,	eicosadienoic acid;
FA,	fatty acid;
γ-LA,	gamma linolenic acid;
G,	gas;
LA,	linoleic acid;
LC,	liquid condensed;
LE,	liquid expanded;
OA,	oleic acid;
π,	surface pressure;
S,	solid;
SDA,	stearidonic acid;

θ,	inclination angle of a monolayer molecule to the A/W interface;
th,	thickness;
UFA,	unsaturated fatty acid;
PUFA,	polyunsaturated fatty acid.

1. INTRODUCTION

There is a plethora of biomolecules that are known to act on membrane level. Their molecular targets can be situated just at the outer cell or deeper – at the cytoplasmic membrane - or within the cytoplasm (Maillard, 2002). In the former case, a biomolecule can partition into the cellular membrane, thus changing its mechanical and physical properties, such as membrane fluidity, which influence cellular functioning. In other cases, a biomolecule has to penetrate through the outer cell wall to reach its target site. The mode of action of such membrane-targeted biomolecules is based on specific interaction with particular intracellular or membrane components. Their identification is of utmost importance as it allows verifying/understanding of their mode of biological activity. Consequently, new biomolecules of enhanced pharmacological activity, better selectivity and decreased toxicity can be designed.

Molecular targets for membranotrophic biomolecules can be selected basing on examining their membrane interactions, using natural systems - either isolated (extracted membranes) or not (living cells) - or artificial membrane models, like Langmuir monolayers, lipid vesicles or liposomes, black lipid membranes, supported lipid bilayers (reviewed in Chan & Boxer, 2007; Eeman & Deleu, 2010). The interplay between biological experiments and artificial models is undoubtedly beneficial. However, some issues are impossible to explain from biological experiments performed on native cell membranes, which are complicated and highly variable structures. In contrast, membrane models, especially Langmuir monolayer (Gaines, 1966), offer the possibility of selective verification of the affinity of a biomolecule to a particular membrane component by examining their mutual interactions in a very systematic way (Maget-Dana, 1999). With Langmuir monolayers one can easily construct a membrane of interest as exemplified in Ref. (Dynarowicz-Latka & Hac-Wydro, 2014). The simplest model would be a one-component monolayer, formed by one of the major membrane lipids (e.g., DPPC) or other selected membrane lipid of interest. Although such a model is certainly an oversimplification, it can be useful in some cases, for example for an easy verification of a particular membrane component, which is important in targeting of a biomolecule to the interior of the cell. More realistic are multicomponent Langmuir monolayers as membrane mimicking systems, prepared by mixing of their representative components in an appropriate proportion found in natural systems. With the monolayer technique it is easy to increase step-by-step the complexity of the model membrane by increasing the number of its components as well as adjusting surface pressure to the value occurring in living cells.

The majority of investigations aimed at investigating biomolecule-membrane interactions are focused on drugs (Nobre et al., 2015). Much less attention has been devoted to dietary supplements. This paper summarizes the results obtained so far for polyunsaturated fatty acids (PUFAs) in artificial membrane environment, modelled with the Langmuir monolayer technique.

2. The Langmuir Monolayers

2.1. Classical Langmuir Monolayer Technique

The technique of monomolecular layers formed at the aqueous solution-air interface, termed the *Langmuir monolayer technique*, is a method in which surface active molecule of an amphiphilic (amphipathic) structure and proper hydrophilic-hydrophobic balance (resulting from the presence of a polar "head", anchoring molecule at the water surface, and an apolar "tail" long enough to prevent the dissolution of the molecule in bulk water) is dropped onto the water surface, where it self-organizes into monomolecular layer, called the Langmuir monolayer (film). For a more detailed description of Langmuir monolayers – their formation and physicochemical characteristic - the readers are referred to surface chemistry books (such as Gaines, 1966; Adamson, 1967; Aveyard & Haydon, 1973; Roberts, 1990; Petty, 1996) or review articles (for example Knobler, 1990; Dynarowicz-Latka, Dhanabalan, & Oliveira, 2001 and references therein). Briefly, in a common monolayer experiment, a known amount of water-insoluble surface active compound is dissolved in volatile organic solvent (most frequently chloroform) and then an aliquot is placed on the water surface using a microsyringe. After solvent evaporation, the molecules cover the entire surface forming a monomolecular layer, which is then compressed with sliding barrier(s), using the so-called Langmuir trough (Figure 1).

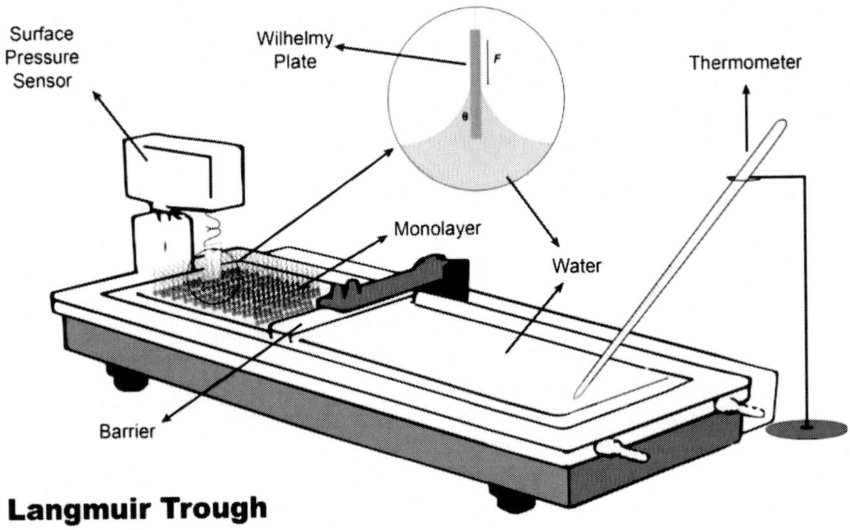

Figure 1. Schematic representation of the Langmuir trough.

As a result, surface pressure - area (π-A) isotherm is recorded, using a Wilhelmy plate as surface pressure sensor. The obtained isotherm for one-component monolayer is characterized by the following parameters (Figure 2):

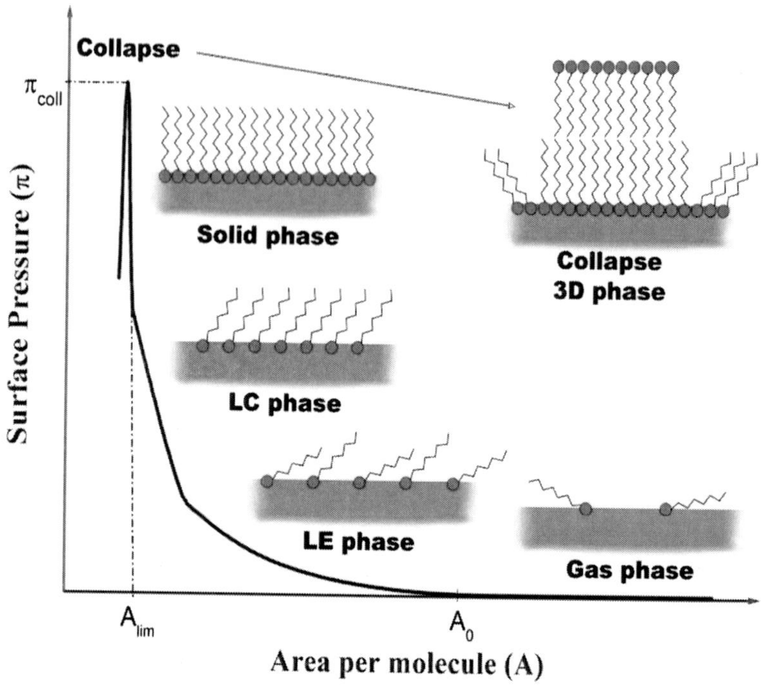

Figure 2. Surface pressure-area π-A isotherm together with characteristic physical states of a monolayer: Gas phase, Liquid-expanded (LE) phase, Liquid-condensed phase (LC), Solid phase.

A_0 – "lift-off" area – which is the molecular area at the surface pressure rise;

A_{lim} – area extrapolated from the steepest part of the isotherm to $\pi = 0$

π_c – collapse pressure – surface pressure at which monolayer collapses

At the monolayer collapse, seen in the course of the isotherm as maximum value of surface pressure to which the monolayer can be compressed (visible as a spike or plateau), two-dimensional (2D) monolayer transforms to three-dimensional (3D) phase.

Physical states of a monolayer upon its compression can be distinguished by calculating the compressibility moduli (C_s^{-1}) values (Davies & Rideal, 1963) (see Table in Figure 3) estimated from the experimental isotherm's datapoints:

Physical state	G	LE	LC	S
C_s^{-1}	0 – 12.5	12.5 - 50	100 – 250	> 250

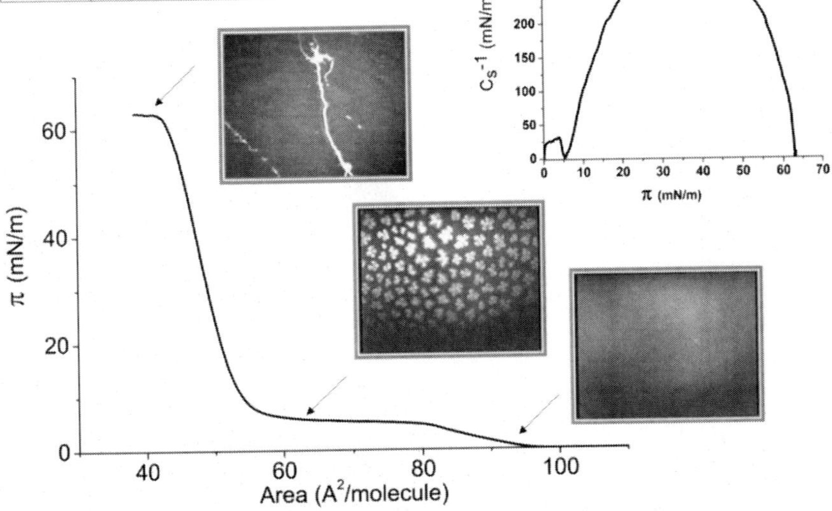

Figure 3. Surface pressure-area (π-A) and compressibility modulus-surface pressure (C_s^{-1}-π) dependencies for DPPC monolayer (Miñones Conde, 2001).

$$C_s^{-1} = -A\left(\frac{d\pi}{dA}\right)_T \tag{1}$$

By plotting the compressibility modulus values *versus* surface pressure it is possible to identify the physical state of the monolayer at a particular surface pressure value, as exemplified in Figure 3 for a monolayer formed by dipalmitoylphosphatidylcholine (DPPC), a model membrane phospholipid.

In the C_s^{-1} *vs* π curves, the phase coexistence, which is seen as a plateau in the π/A isotherm, appears as a minimum, while the surface pressure at which monolayer collapses can be depicted at the point where C_s^{-1} is reaching 0.

Traditional Langmuir monolayer experiments are usually complemented with microscopic methods; Brewster angle microscopy (BAM) being most frequently used. This method enables direct visualization of the monolayers structure at particular stages of compression (as exemplified for DPPC monolayer, Figure 3) as well as film thickness measurements, as described below.

2.2. Visualization of Monolayers' Texture - Brewster Angle Microscope (BAM)

In 1990, based on research from Sir David Brewster, two scientific groups (Henon & Meunier, 1991) (Hönig & Möbius, 1991) simultaneously developed a powerful tool that enabled visualization of the monolayer morphology: Brewster's Angle Microscope (BAM). Although there is also another microscopic technique that allows direct visualization of monolayers structure - the fluorescence microscopy (Moore, Knobler, Broseta, & Rondelez, 1986) - however, BAM is advantageous since it does not require the introduction of any additives (fluorescent probe) that may interfere with film molecules, which decreases the risk of experimental artefacts.

This method involves the emission of a laser beam with *p*-polarized light over the air-water interface at an angle equal to 53.15 (i.e., Brewster's angle). Under this angle of incidence, no light is reflected from the interface between two media of different refractive indices (Figure 4). The introduction of a Langmuir monolayer modifies the Brewster angle conditions, and light is now reflected and partially polarized, which is used for imaging of the surface structure (see Figure 3, BAM images for DPPC monolayer). In a quantitative application of Brewster angle microscopy, one can measure the reflectivity (R), applying the relationship between the intensity of the reflected light (I) and the incident light (I_0): $R = I/I_0$.

The estimation of the monolayer's thickness requires a more detailed analysis (Azzam & Bashara, 1992). Namely, the intensity of light reflected under the Brewster angle depends on the thickness and optical properties of the film, as indicated by the equation:

$$I = |Rp|^2 = C \cdot d^2 \quad (2)$$

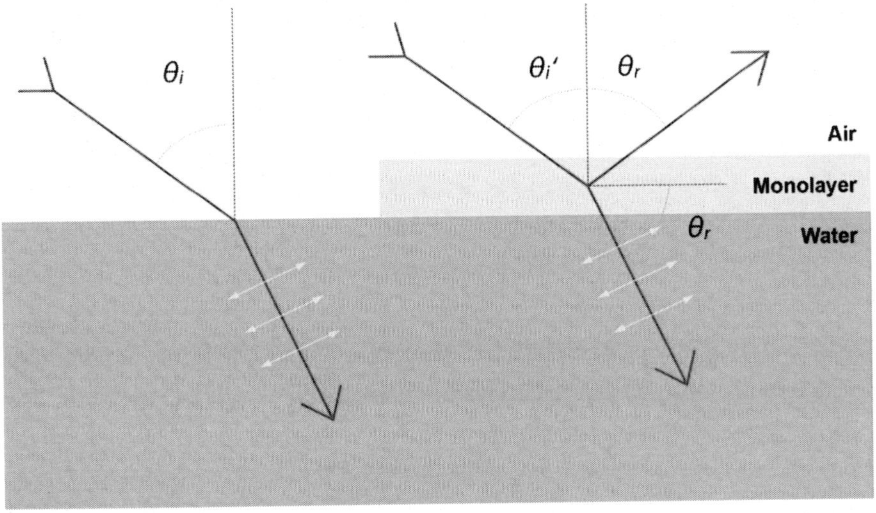

Figure 4. Reflected light at the Brewster angle at the air/water interface (left) and with a monolayer (right) (Fidalgo Rodríguez, 2016).

where *I* is the relative reflectivity, *C* is a constant; *d* is the thickness of the monolayer and *Rp* is the reflectivity of the p-polarized reflected beam, which depends on a set of optical parameters, such as wavelength of incident light, refractive indexes of air, monolayer and water and angles of incidence of light relative to planes of incidence of air, monolayer and water. Some of these parameters are usually unknown, especially the refraction index of the monolayer, but this does not prevent us from knowing the *relative* thickness of the film in a certain surface state in relation to a different one (de Mul & Mann, 1998), although theoretical models have been developed to calculate the actual thickness values (Overbeck, Hönig, Wolthaus, Gnade, & Möbius, 1994) (Overbeck, Hoenig, & Moebius, 1993) (Weidemann, Gehlert, & Vollhardt, 1995).

The Brewster angle microscope has an optical-mechanical unit that consists primarily of two parts:

1. *Laser arm.* It contains the laser, which emits at a specific wavelength radiation and an adjustable power. In this arm, there is also a polarizer that ensures that the beam is p-polarized with respect to the plane of incidence.

2. *Scanner arm*. It contains three components:

a) *CCD camera*: Converts the intensity of the reflected light by the sample into a video image, which can be observed on the monitor. These images have a resolution of 2 μm.

b) *Analyzer*: It allows the collection of additional information about the anisotropy of the sample, by rotating this component it is possible to distinguish anisotropic domains from other contrasts caused by differences in density or thickness in isotropic films.

c) *Scanner*: It consists of a mechanical device that allows moving the lens along its optical axis in order to focus all the areas of the image, so that the final image is obtained from the different scanned areas during its movement.

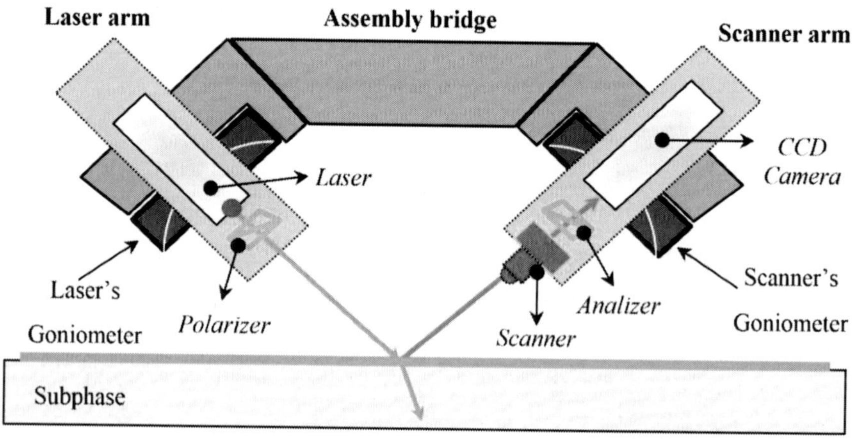

Figure 5. Scheme of Brewster angle microscope.

2.3. Interaction between Molecules in Multicomponent Monolayers

In multicomponent monolayers, interactions between film molecules can be analyzed both qualitatively and quantitatively (Dynarowicz-Latka & Kita, 1999). The methodology is based on calculating the excess parameters of mixing, which depict deviations of the investigated system from ideality. In the ideal system of two components (1 and 2), mean molecular area in mixed monolayer (A^{id}_{12}) can be calculated as a sum of molecular areas of each component (A_1, A_2), determined from experimental isotherms for one-component monolayers, multiplied by respective values of their mole fractions (X_1, X_2):

$$A_{12}^{id} = x_1 A_1 + x_2 A_2 \tag{3}$$

This dependence is linear for systems, where monolayer components mix ideally (i.e., interactions between molecules 1-2 are the same as between molecules 1-1 and 2-2) or when components do not mix at all. Any deviations from ideal behavior is depicted as non-linearity and can be qualified by the excess area of mixing values (A^{exc}), defined as:

$$A^{exc} = A_{12} - A_{12}^{id} = A_{12} - (x_1 A_1 + x_2 A_2) \tag{4}$$

where A_{12} is the experimental mean molecular area for experimental mixed monolayer.

With the A^{exc} *versus* $X_{1(2)}$ plot it is possible to examine in a *qualitative* way the kind of deviations from ideality (positive or negative). Negative values indicate that interactions between components in mixed monolayer (1-2) are more attractive (or less repulsive) compared to interactions between molecules in one-component monolayers (1-1; 2-2). On the other hand, positive values indicate that the interactions in the mixed monolayers are less attractive (or more repulsive) than in the corresponding one-component monolayers. The case of A^{exc} = zero is a proof that the system behaves ideally.

Collapse pressure (π_{coll}) is also an important parameter since - by plotting π_{coll} values as a function of film composition - one can verify whether the investigated monolayer components mix in monolayer or not. Namely, according to the phase rule, if two components are miscible, the collapse pressure varies with mixed film composition (Crisp, 1949) (Wu &

Huntsberger, 1969), while for immiscible system, the collapse pressure remains constant.

The interactions are usually *quantified* with the excess thermodynamic functions of mixing, most frequently with the excess free enthalpy of mixing, ΔG^{exc}, defined as (Pagano & Gershfeld, 1972):

$$\Delta G^{exc} = N_A \int_0^{\pi} A^{exc} \, d\pi \qquad (5)$$

wherein N_a is the Avogadro number.

From these values, the strength of interaction between molecules can be estimated. Negative (positive) values of ΔG^{exc} indicate that attractive (repulsive) interactions in mixed monolayer are stronger as compared to the respective monolayers formed by pure components. Different approaches used to characterize molecular interactions in monolayers are reviewed in more details in paper by Dynarowicz-Latka & Kita, 1999, while the paper by Gong, Feng, Go, & Soew, 2002, provides the readers with useful experimental examples of the analysis of intermolecular interactions using the Langmuir monolayer technique.

2.4. Langmuir Monolayers as Membrane Models

Langmuir monolayers are the oldest cell membrane models ever reported in the literature (Bagatolli, Ipsen, Simonsen, & Mouritsen, 2010). Already in 1925, Gorter & Grendel compared, using the Langmuir through, the area occupied by lipids extracted from erythrocythes, with the average surface area of living red blood cells. This simple experiment evidenced that the cell membrane is composed of lipid bilayer.

The application of the Langmuir monolayer technique to model cellular membranes comes from the fact that the membrane components are amphiphilic in structure and are surface active, which enables them to self-organize on the air/water interface and form Langmuir monolayers. Therefore, Langmuir films formed by membrane components (such as phospholipids, glycolipids, sterols) may serve as simple, physical models of biomembranes. Although natural membrane is a bilayer, and the Langmuir monolayer represents its one leaflet, however, correlation has been found between monolayers and bilayers. Namely, due to so-called monolayer-bilayer correspondence (Brockman, 1999, Feng, 1999), monolayers reveal similar properties (such as area per molecule, phase transition, elastic compressibility) to bilayers at surface pressures of 30-35 mN/m. Therefore, at these conditions, Langmuir monolayer can successfully mimic natural bilayer system. The advantage of using Langmuir monolayer as a biomembrane model is its simplicity of preparation and possibility of a continuous, easy and precise control of both quality of the surface and such parameters as molecular packing, physical state, lateral pressure and composition. By mixing the membrane components in a certain proportion it is possible to prepare various types of tissues/cells membranes or individual membrane leaflets. Additionally, by mixing membrane lipids and bioactive molecules in a monolayer, it is possible to study their mutual interactions and examine their effect on model membrane. Recent articles by Stefaniu, Brezesinski, & Möhwald, 2014 and Nobre et al., 2015, provides an up-to-date review of successful applications of the Langmuir monolayer technique in drug-membrane interaction studies. However, a basic characteristics of interactions, based solely on surface pressure measurements, is usually not sufficient for a thorough characterization of the investigated system and requires additional analysis with modern surface-sensitive techniques (for a review see Dynarowicz-Latka et al., 2001, such as optical (Brewster angle microscopy, BAM; fluorescence microscopy), diffraction (XRD) and linear and nonlinear spectroscopic (PM-IRRAS, SERS, SHG, SFG) methods. Theoretical studies based on simulation methods provide also an important contribution to the thorough characteristics of bioactive molecules and their membrane interactions.

3. PUFAs in Langmuir Monolayers

3.1. PUFAs at the Air/Water Interface

Physicochemical properties of Langmuir monolayers formed by unsaturated fatty acids have a long history of studies. First reports showed that oleic acid (18:1), having one double bond in apolar chain, formed much more expanded surface films as compared to condensed monolayers from saturated stearic acid (18:0) molecules, having the same chain length (Langmuir, 1917; Adam & Jessop, 1922). Further research in this respect, conducted by Hughes (Hughes, 1933), confirmed earlier assumptions that the presence of unsaturated bond(s) in the fatty acid structure exerts a significant effect on the molecular organization in Langmuir monolayer, influencing its physical state, which results in significant film expansion. In subsequent studies (Marsden & Rideal, 1938), the influence of the type of *cis* and *trans* isomerization on the surface area occupied by monounsaturated fatty acid molecules in Langmuir monolayers has been examined. Results of these studies have clearly demonstrated that unsaturated fatty acids having a double bond in the *trans* conformation form condensed monolayers, similarly to saturated fatty acids of the same chain length, while unsaturated fatty acids possessing a double bond in the *cis* conformation form expanded surface films. Later, a more detailed study was performed (Schneider, Holman, & Burr, 1949), which proved that in naturally occurring unsaturated fatty acids (having a double bond in the *cis* conformation), the area per molecule increase correlates with the number of unsaturated bonds in the hydrophobic chain. In addition, difference in the limiting area between *all-cis* and *all-trans* isomers of polyunsaturated fatty acids was found to depend whether the molecule has a conjugated system of double bonds (i.e., eleostearic acid - 18:3), or not (α-linolenic acid - 18:3). In the former case, the observed effect was smaller.

Subsequent papers on polyunsaturated fatty acids in Langmuir monolayers were focused on studying the influence of pH on monolayers characteristic and film stability as well as determination of their surface dissociation constant (Tomoaia-Cotişel, Zsako´, Mocanu, Lupea, & Chifu,

1987); influence of metal ions present in water subphase in order to increase their stability for their transfer onto solid supports (Peltonen & Rosenholm, 1989), and – upon discovery of the modern physicochemical surface-sensitive techniques (such as BAM and GIXD) – the effect of *cis* and *trans* isomerization on the phase behavior, domain morphology and two-dimensional lattice structure of PUFAs monolayers (Vollhardt, 2007). The obtained results were in agreement with those of former studies, proving that unsaturation causes the monolayer expansion, the extent of which is larger for the *cis* as compared to the *trans* isomers. The thermodynamic analysis done on the basis of Langmuir isotherms showed that the presence of unsaturated bond in the hydrophobic chain of the fatty acid molecule caused similar effect as shortening of the alkyl chain length (referred to as the "hydrophilic effect"), and its extent was again more pronounced for the *cis* versus *trans* isomers. Moreover, BAM studies of the films structure at different temperatures revealed significant differences in the condensed domains structure appearing at the LE-LC phase transition in *cis* and *trans* isomers, which confirmed previous studies (Iimura, Yamauchi, Tsuchiya, & Kato, 2001) showing the formation of dendritic *versus* round-shaped domains for *cis* and *trans* isomers, respectively, as illustrated in Figure 6.

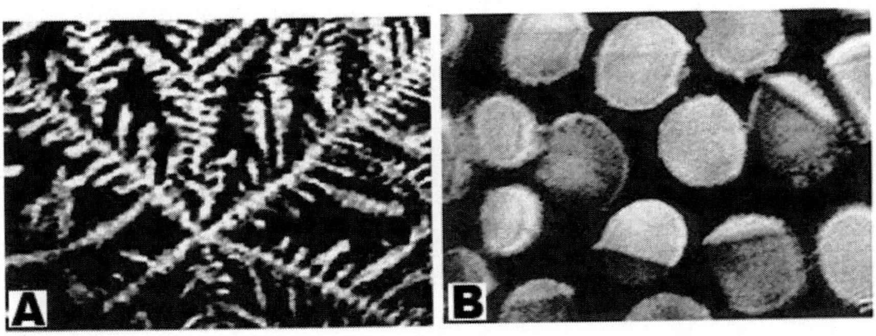

Figure 6. Shape of condensed domains observed with BAM at the LE-LC transition in Langmuir monolayers of erucic (cis-13-docoseoic) acid (A), and elaidic (trans-9-octadecanoic) acid (B). Reproduced with permission from (Iimura et al., 2001).

Recently, systematic Langmuir monolayer studies of PUFAs have been published (Fidalgo Rodríguez, Dynarowicz-Latka, & Miñones Conde,

2017). The studied unsaturated fatty acids have been examined in ω-3, ω-6 and ω-9 series, which are of biological importance (according to the nomenclature by Holman, 1964, for omega-x fatty acids, where x indicates the location of the last double bond counting from the last carbon atom). Since it has been proved that the unsaturation in the hydrocarbon chain causes the increase of water solubility of fatty acids (Tomoaia-Cotişel et al., 1987), in order to minimize this effect upon comparing the results, the measurements have been performed at low subphase temperature (10 °C). The most important results are shown below.

3.1.1. The Influence of Acyl Chain Unsaturation

PUFAS monolayers with increasing degree of unsaturation in their alkyl chains (2, 3 and 4 double bonds) spread on pure water subphase at 10 °C exhibit π-A isotherms with similar characteristics (Figure 7). At the beginning of compression, all the monolayers show a gas-liquid expanded (G-LE) transition, where the surface pressure remains constant showing a "plateau" with π values around zero mN/m. As the area decreases by the compression of the monolayers, a monotonic increase in surface pressure was observed during compression indicating that the monolayers remain in the liquid-expanded (LE) state. The limiting areas (A_0), obtained by extrapolation of the steep linear segment of the π-A curves to $\pi = 0$ mN/m, increase with the number of double bonds located in the hydrocarbon chain possessing the same number of carbon atoms (see area values in Table 1 for linoleic acid (LA 18:2, ω-6), α-linolenic acid (α-LA 18:3 ω-3) and stearidonic acid (SDA 18:4, ω-3) monolayers).

These A_0 values are similar to those reported in literature (Tomoaia-Cotişel et al., 1987) (Linden & Rosenholm, 1995) (Peltonen & Rosenholm, 1989) (Makyła & Paluch, 2009), although no references to SDA monolayer has been found.

Monolayers of liquid surfactants in bulk at room temperature collapse at a characteristic surface pressure, π_{coll}, which is close to their equilibrium spreading pressures (ESP) (Gaines, 1966). The values of π_{coll} compiled in Table 1 prove that such a behaviour is also characteristic for the investigated PUFAs. Indeed, π_{coll} value for LA monolayer (28.7 mN/m) is close to its ESP, which is 29 mN/m (Linden & Rosenholm, 1995). On the other hand, the value of π_{coll} for a monolayer is the evidence of its stability; i.e., the higher this value is, the more stable monolayer is. From results in Figure 7 and data of Table 1, it is evident that the π_{coll} values of the studied PUFAs decrease with the number of double bonds in the hydrocarbon chain, which means that monolayers become less stable upon increase of the unsaturation degree of the aliphatic chain.

Table 1. Characteristic parameters corresponding to PUFAs monolayers: "lift-off" area (A_0) limiting area (A_{lim}), collapse pressure (π_{coll}) and maximum value of compressibility modulus (C_s^{-1}) (Fidalgo Rodríguez, 2016)

PUFA	A_0 (Å²/molec)	A_{lim} (Å²/molec)	π_{coll} (mN/m)	C_s^{-1} max (mN/m)
Linoleic acid (18:2) ω-6	67	46.7	28.7	95.2
α-Linolenic acid (18:3) ω-3	85	49.8	27.3	72.4
γ-Linolenic acid (18:3) ω-6	80.5	55.1	27.9	72.8
Stearidonic acid (18:4) ω-3	106.6	62.7	25.8	54
Eicosadienoic acid (20:2) ω-6	74.5	50.1	28.1	95.5

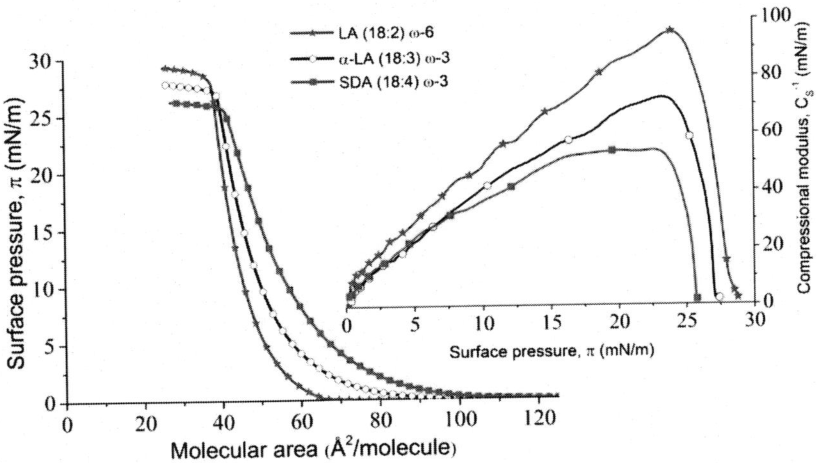

Figure 7. π-A isotherms for linoleic acid (LA), α-linolenic acid (α-LA) and stearidonic acid (SDA) monolayers spread on water at 10 °C. C_s^{-1}-π plots in the inset (Fidalgo Rodríguez, 2016).

A more accurate analysis of the physical state of the investigated films is obtained from the plot of the compressibility modulus values (C_S^{-1}) as a function of surface pressure (π). The compressibility modulus was calculated from the experimental data of the π-A isotherms. The maximum C_S^{-1} values in the C_S^{-1}- π curves for the studied monolayers (inset of Figure 7) are smaller than 100 mN/m, which correspond to monolayers in liquid-expanded state (Davies & Rideal, 1963). At high pressures, the compressibility curves exhibit minima values (C_S^{-1} = 0), which indicate their collapse. As it can be seen, these minima (π_{coll}) decrease with the increased degree of unsaturation of the hydrocarbon chains, proving their lower stability, as commented above. From the results in Figure 7 it can be concluded that the increase of unsaturation in the hydrocarbon chains causes a decrease of C_S^{-1} values, i.e., monolayers adopt a more expanded surface state. Because the studied unsaturated fatty acids have the same polar (carboxylic) group anchored in bulk water phase, the observed differences in monolayers behaviour obviously arise from divergences in apolar moieties of these molecules, which determine their respective structures, i.e., the molecular conformation of the acyl chains in their monolayers must be different, according to the number of double bonds.

3.1.2. The Influence of the Double Bonds Position

The position of the double bonds in the acyl chains of PUFAs has little influence in the behavior of their monolayers. Indeed, α-LA (18:3, ω-3) and γ-linolenic acid (γ-LA 18:3, ω-6) monolayers exhibit similar π_{coll} values: 27.3 mN/m and 27.9 mN/m, respectively (Figure 8 and Table 1), which indicates that the stability of both films is practically unaffected by the different position of the double bonds in the hydrocarbon chains of PUFAs having the same number of carbon atoms. In the same way, the position of the double bonds in the acyl chains hardly affects the compressibility of monolayers: α-LA forms a fluid film ($C_S^{-1}{}_{max}$ = 72.4 mN/m), similarly to γ-LA mononolayer ($C_S^{-1}{}_{max}$ = 72.8 mN/m). On the other hand, both C_S^{-1}- π curves overlap at pressures above 20 mN/m (inset of Figure 8). The greatest impact of the double bond position on the conformation of the PUFAs molecules at the air/water (A/W) interface is observed in the limiting area values of both compounds: α-LA exhibits a lower limiting area (49.8 Å²/molecule) *versus* the corresponding γ- LA fatty acid (55.1 Å²/molecule). This behaviour was attributed to a higher water solubility of the α-LA monolayer relative to the γ- LA one (Fidalgo Rodríguez et al., 2017). Indeed, the film molecules loss from the interface is the result of their improper hydrophobic/hydrophilic balance. Because these PUFAs have the same (carboxyl) polar headgroups, it can be assumed that their hydrophilic character is the same (at a particular temperature). However, the hydrophobic character of the acyl chains depends on the double bonds position, in addition to their length and unsaturation degree. According to this, the γ-LA hydrophobicity must be greater (lower water solubility) than that of α-LA monolayer. These differences of hydrophobicity must result from a different molecular configuration of both compounds, as it will be discussed later.

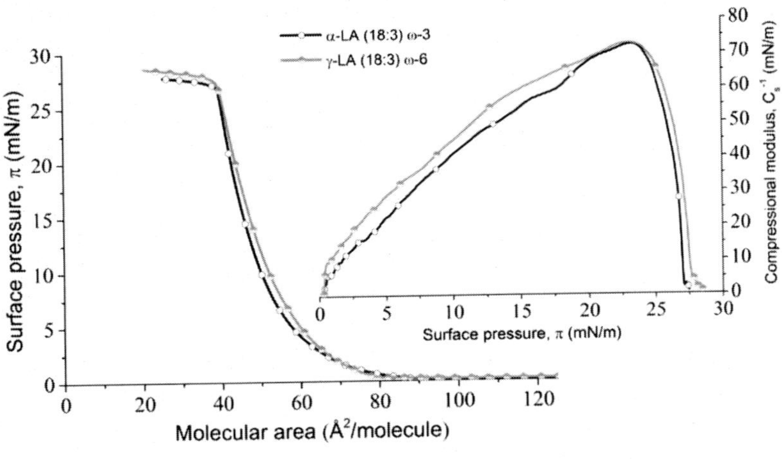

Figure 8. π-A isotherms for α-LA and γ-LA monolayers spread on water at 10 °C. C_s^{-1}-π plots in the inset. (Fidalgo Rodríguez, 2016).

3.1.3. Effect of the Acyl Chain Length

The increase of the hydrocarbon chain length in PUFAs with the same number of double bonds (LA, 18:2, ω-6 and eicosadienoic acid, EA 20:2, ω-6) causes an increase of the film areas per molecule (Figure 9). The A_o values differ from 46.7 Å²/molecule for LA monolayer to 50.1 Å²/molecule for EA film (Table 1). These differences were attributed to the increased hydrophobicity of the EA aliphatic chains as a result of its longer length, causing its decreased water solubility (J. L. Fidalgo Rodríguez et al., 2017). However, the number of carbon atoms in the alkyl chain hardly influences the compressibility of monolayers, since LA as EA monolayers have quite a comparable C_S^{-1} values along the film compression (inset of Figure 9). In the same way, the addition of two carbon atoms to the hydrocarbon chain of LA, leading to formation of EA, hardly influences the value of the collapse pressure of EA in relation to that of the LA monolayer (Table 1).

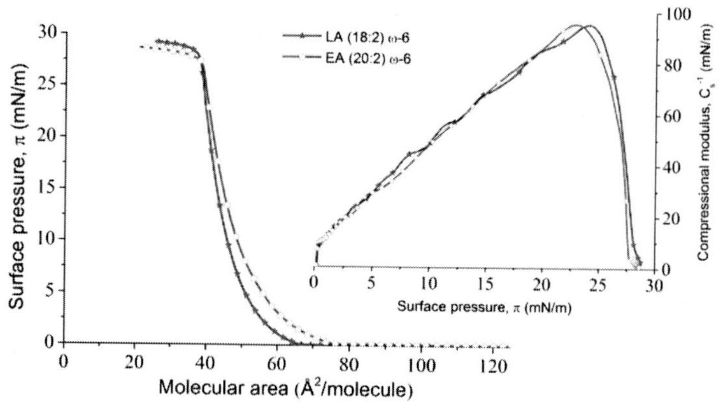

Figure 9. π-A isotherms for LA and EA monolayers spread on water at 10°C. C_s^{-1}-π plots in the inset (Fidalgo Rodríguez, 2016).

3.1.4. Monolayers Thickness and Molecular Configuration

In order to get insight into molecular conformation of PUFA monolayers at different surface states along the film compression, monolayers thickness (th) measurements were recorded and compared to the results for mono-unsaturated fatty acid: cis-oleic acid - OA (18:1). In all the cases, at the beginning of compression, when monolayers exhibit the G-LE phase transition, the average value of the films thickness remains constant with compression, similarly to the surface pressure. In this region, except for EA monolayer, the film thickness values ($th_{G\text{-}LE}$) decrease with increasing the number of double bonds in the alkyl chain (Figure 10 and Table 2).

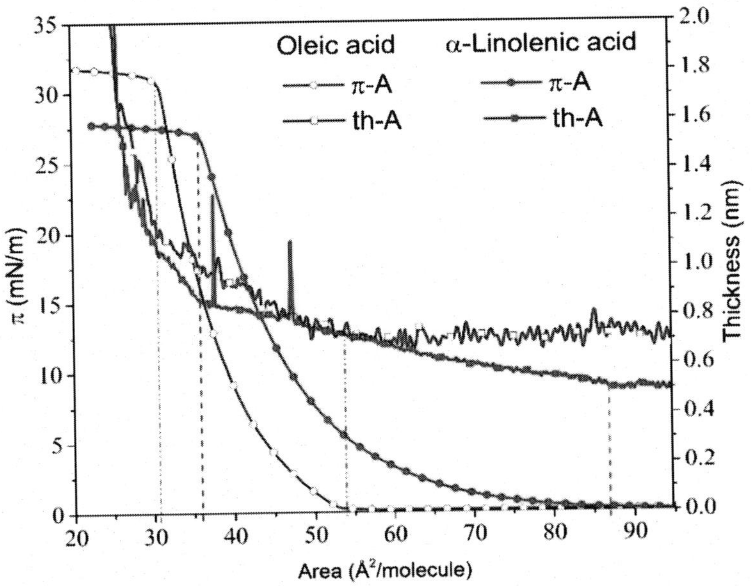

Figure 10. π-A isotherms and th-A for oleic (blue line - open symbols) and α-linolenic acid (red line-filled symbols) monolayers spread on water at 10°C (Fidalgo Rodríguez, 2016).

Table 2. Values of characteristic parameters corresponding to UFAs monolayers: distances between C atoms, monolayers thickness and tilt angles of hydrocarbon chains (Fidalgo Rodríguez, 2016)

Fatty acid	Theoretical distance, d (nm)	Thickness (nm)		Tilt angle
		G-LE	Collapse	Collapse (θ)
Linoleic acid (18:2) ω-6	1.13 (C1-C12)	0.69	1.08	19°
α-Linolenic acid (18:3) ω-3	0.85 (C1-C10)	0.51	0.86	24.1°
γ-Linolenic acid (18:3) ω-6	0.78 (C1-C9)	0.47	0.78	24.9°
Stearidonic acid (18:4) ω-3	0.83 (C1-C10)	0.49	0.83	31.4°
Eicosadienoic acid (20:2) ω-6	1.36 (C1-C14)	0.85	1.33	15.8°

In the LE region, all the recorded th-A curves exhibit similar shape, and the monolayer thickness increases slowly and linearly with the film compression. Interestingly, the molecular area, at which the increase of thickness begins, coincides with the lift-off area corresponding to the increase of surface pressure (Figure 10). Thus, at this molecular area value, it can be postulated that the fatty acid molecules undergo a change in orientation from a position more or less tilted with respect to the surface of the water in the LE-LC transition region to an increasingly vertical position in the LE phase. Finally, when monolayer collapses, a sudden increase in film thickness takes place, due to multilayers formation.

The more stable molecular conformations for the studied unsaturated fatty acid (UFA) molecules listed in the US National Center for Biotechnology Information (NCBI), are shown in Figure 11. For all of them, the theoretical distance (*d*) (obtained from data of NCBI) between the carboxyl carbon atom anchored into water subphase (denoted as C_1 atom) and the most distant carbon atom from the A/W interface (C_{dist}), virtually coincides with the experimental thickness values at the film collapse (th_{coll}) (Table 2). Thus, in the case of OA monolayer, the carbon atom number 12 is the most distant from anchor location (Figure 11), being d = 1.17 nm, a value that practically coincides with the experimental thickness at the collapse point (1.15 nm) (Figure 11 and Table 2). For LA, the carbon atom being most distant from the water surface is also C12, but in other molecules it is different, as indicated in the table: C10 in α-LA and SDA, C9 in γ-LA and C14 in EA. For all the studied compounds, these most distant carbon atoms, which determine the maximum thickness of their monolayer, are in vertical position with the carboxyl carbon anchored into the water subphase. However, due to the unsaturation, the aliphatic chains preceding these distant atoms are not vertical, but tilted in respect to the surface normal (Figure 11). It is possible to calculate the tilt angle (θ) of these tails using, on one hand, the theoretical *d* values and, on the other, the length of the hydrocarbon chains preceding the first double bond (for the computation of this acyl chain length it was considered that the distance between neighbouring C-C atoms is 1.53 Å and the angle between them equals 112° (DeWolf, Leporatti, Kirsch, Klinger, & Brezesinski,

1999), although the NCBI program provides directly this length. From these data, a simple trigonometric calculation enables to estimate the tilt angle, θ. The obtained results are shown in Figure 11 and Table 2, where it can be seen that θ values increase with the number of double bonds (except for EA), i.e., the aliphatic chains are systematically more tilted (in respect to surface normal), resulting in the formation of more expanded and more fluid monolayers, as evidenced from the π-A and C_s^{-1}-π plots, described above.

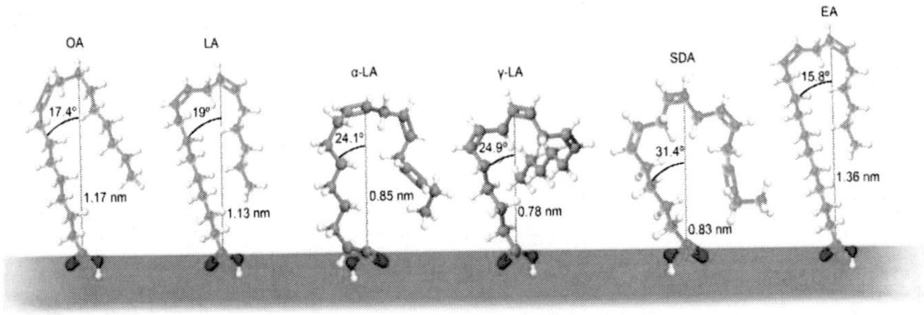

Figure 11. Molecular conformations for PUFAs monolayers at the collapse (Fidalgo Rodríguez, 2016).

The molecular conformation proposed for the hydrocarbon chain of the OA molecule at the A/W interface (Figure 11) allows packing of the two arms of the "hook" when the monolayer is compressed, thus increasing the attractions between chains. However, in the LA monolayer, the existence of an additional double bond increases the rigidity of the aliphatic chain, hindering the close packing of the acyl chains, leading to the formation of a fluid film (Figure 7). Also, the aliphatic chains in EA monolayer are packed similarly to LA, since the two additional carbon atoms in the EA molecule are located in the chain preceding the double bonds, not influencing the molecular packing, but only its length. This means that these two additional atoms affect only the monolayer thickness, but not its fluidity. Moreover, upon increasing of the chain length, the hydrophobic character of the molecule also increases, resulting in its lower solubility. On the contrary, for unsaturated fatty acids with more than two double bonds (α-LA, γ-LA and SDA), the aliphatic chains show a highly coiled conformation (Figure 11), which hinders the close packed arrangement of the tails, leading to more expanded monolayers (occupying larger molecular areas). Furthermore, the shortening of the alkyl chains length, as a result of the coiled conformation, causes a decrease in the hydrophobicity of the molecules, by favoring their water dissolution. The comparison of the molecular conformations of α-LA and γ-LA fatty acids monolayers at the A/W interface in Figure 7 evidences a more coiled and closer packing of the hydrocarbon chains of the γ-LA monolayer, justifying its higher hydrophobicity and lower compressibility (at $\pi<20$ mN/m) as compared to that of the α-LA monolayer.

3.2. PUFAs in Lipid Monolayers

The results described above unambiguously prove that PUFAs are capable of forming monolayers at the free water surface and therefore are suitable for studies with the Langmuir monolayer technique. Subsequent investigations were focused mainly on studying mixtures of unsaturated fatty acids with particular membrane lipids (cholesterol, phospholipids) as well as their mixtures, modeling natural membranes, in order to get insight into hypocholesterolemic effect *in vitro*.

3.2.1. Binary Mixed Monolayers of PUFAs/Membrane Lipid

The simplest membrane model is one-component film formed by a chosen membrane lipid. Since the biological membrane is composed mainly by phospholipids and sterols, in the first approximation, the interactions between fatty acids and a simplified model membrane, composed by either cholesterol or selected phospholipid, have been investigated. Although such an approach is certainly an oversimplification, as natural biomembranes are known for their high complexity, however, such investigations are necessary for interpretation of the results carried out on more complicated membrane-mimicking systems.

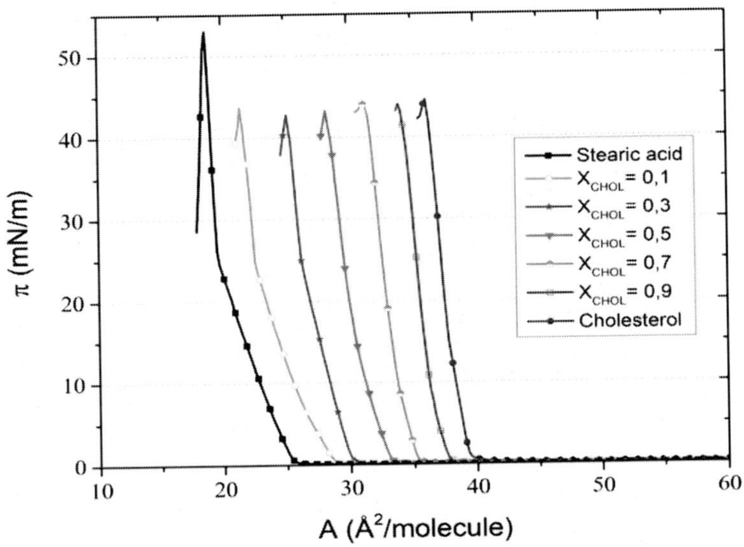

Figure 12. π-A isotherms for pure and mixed monolayers of stearic acid and cholesterol (Fidalgo Rodríguez, 2016).

Firstly, a detailed analysis of interactions in two-component monolayers was carried out between main membrane sterol – cholesterol - and the following unsaturated fatty acids: oleic acid (18:1), linoleic acid (18:2), α and γ-linolenic acid (18:3), stearidonic acid (18:4) and arachidonic acid (18:4) as well as for a saturated – stearic acid (18:0) – for the purpose of comparison (Fidalgo Rodríguez, 2016; Seoane, Dynarowicz-Latka, Miñones Jr, & Rey-Gómez-Serranillos, 2001). The results, based on the analysis of interaction parameters derived from the π/A isotherms, clearly showed significant differences in the properties of mixtures of cholesterol with saturated *versus* unsaturated fatty acids. The presence of one collapse in the course of the isotherms for mixtures of stearic acid/cholesterol, the value of which was found to be identical as that for pure cholesterol (44.2 mN/m) and did not change with mixed film composition, proved that cholesterol and stearic acid did not mix in monolayers (Figure 12).

On the other hand, mixtures of cholesterol and unsaturated fatty acids were found to behave differently. Indeed, Figure 13, which corresponds to the mixed system formed by α-linolenic acid and cholesterol - as an example of PUFAs – sterol mixed films- shows that for mixtures containing X_{chol}= 0.1-0.3, the component which collapses more easily (α-LA) is ejected from the mixed monolayer at surface pressures close to the value characteristic for one-component monolayer from the investigated PUFA. The application of the phase rule (Crisp, 1949) indicates that cholesterol and α-LA mix in these binary films only in the surface pressure region below π_{coll} of the PUFA. Further increase of surface pressure after the collapse is due to the presence of cholesterol molecules, which collapse is not recorded in the π-A isotherms as a consequence of its low content in the mixtures. However, for mixtures of X_{chol}= 0.5-0.9, the collapse occurs at surface pressures slightly higher than that for cholesterol, indicating high stability of these monolayers at low and high surface pressures.

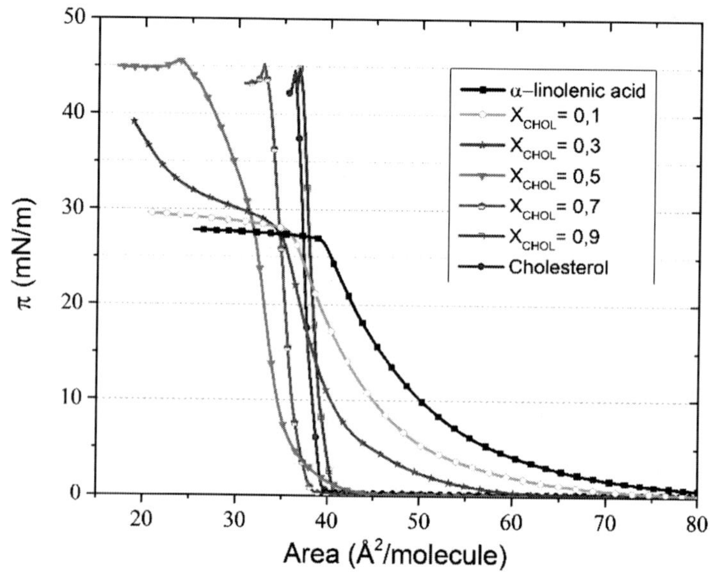

Figure 13. π-A isotherms for pure and mixed monolayers of α-linolenic acid and cholesterol (Fidalgo Rodríguez, 2016).

Negative ΔG^{exc} values at surface pressures ranging from 2.5 mN/m to 25 mN/m for all the studied unsaturated fatty acids binary systems with cholesterol proved that these mixtures are non-ideal and interacting (Figure 14).

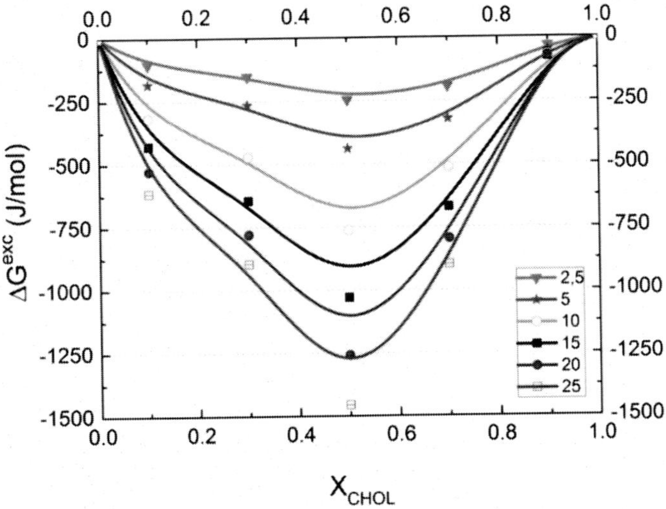

Figure 14. Excess Gibbs energy of mixing (ΔG^{exc}) versus molar fractions (X_{chol}) in α-LA/Chol mixed monolayers at different surface pressures (Fidalgo Rodríguez, 2016).

These results confirm that the attractive interactions in the binary films are stronger and more energetically favorable than those in the one-component films. Besides, the negative values of ΔG^{exc} are lowered with the increasing surface pressure which implies that the intermolecular attractive interactions are stronger when the mixed monolayer is in a more condensed state, where the molecules are found to be at shorter distances and the van der Waals forces between the hydrocarbon tails are greater. On the other hand, the maximum negative values of ΔG^{exc} (maximum film stability) were found in all the studied mixed films for, approximately, the equimolar mixtures, which corresponds to the possible formation of stable surface complexes of 1:1 stoichiometry. Also, interestingly, it was found that the strength of interactions depends on the kind of unsaturated fatty acid, i.e., the degree of unsaturation and the position of the double bonds in the acyl chain. The bar graph in Figure 15 shows the following features: *a)* for the saturated stearic acid - cholesterol system, $\Delta G^{exc} \approx 0$, because both components are immiscible at the A/W interface (the collapse surface pressure of the different mixtures remains constant upon varying films composition); *b)* the attractive interactions ($\Delta G^{exc} < 0$) between cholesterol and unsaturated fatty acids increase with increasing of the alkyl chains unsaturation until a maximum value of 3 double bonds; *c)* for the same degree of unsaturation, the fatty acid - cholesterol interaction is greater if the fatty acid belongs to the omega-3 than to other series (the negative value of ΔG^{exc} in the graph is higher for the mixed film of α-LA/Chol than for γ-LA/Chol).

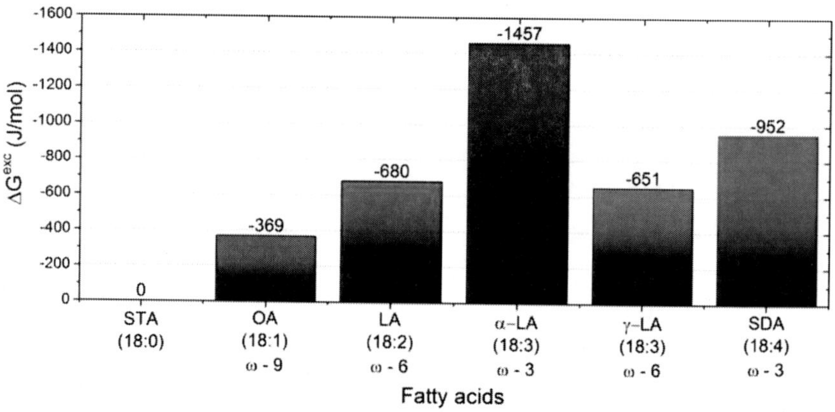

Figure 15. ΔG^{exc} values at 25 mN/m for equimolar mixed films of fatty acids and cholesterol (Fidalgo Rodríguez, 2016).

In another work, a monolayer formed by either saturated (distearoylphosphatidylcholine, DSPC) or unsaturated (dioleoyl-phosphatidylcholine, DOPC) phospholipid was applied as simple biomembrane model, to which selected unsaturated fatty acids were introduced: oleic acid, linoleic acid and - for comparison - stearic acid (Hąc-Wydro, Jędrzejek, & Dynarowicz-Latka, 2009). The results confirmed that the influence of the studied fatty acids on the model membrane is closely related to the degree of unsaturation of the hydrocarbon chain, reflecting in the geometrical conformation of the molecule. The introduction of saturated fatty acid (of a rod-like structure) into such a model membrane increased its rigidity, while the incorporation of unsaturated fatty acid (of the coiled conformation) made it more fluid. Moreover, it has been shown that the studied unsaturated fatty acids interact stronger with saturated (DSPC) than unsaturated (DOPC) phospholipid molecules and exert a stronger condensing effect on DSPC *versus* DOPC monolayer. The effect of stearic acid was significantly less pronounced. It has also been shown that the introduction of even a small amount of fatty acids into mixed DSPC/DOPC model membrane destabilized the system.

3.2.2. PUFAs in Multicomponent Monolayers Mimicking Cell Membranes

Further research was aimed at determining the effect of unsaturated fatty acids on the physicochemical properties of multicomponent artificial membranes modeled with Langmuir monolayer techique (Hac-Wydro & Wydro, 2007). A two-component mixture, consisting of cholesterol and a model phospholipid: DPPC, mixed in the proportion of 0.5 (characteristic for the majority of human membranes) was adopted as a biomembrane model. Then, the following fatty acids: oleic acid and α-linolenic acid as well as stearic acid (for comparison) were added to the mixed lipid monolayer, modeling a biomembrane. Binary mixtures of fatty acid/cholesterol and fatty acid/DPPC have also been examined to enable the interpretation of the ternary fatty acid/cholesterol/DPPC systems. The most important finding is that stearic acid increases the rigidity of model membrane, which becomes less stable from thermodynamic point of view, while unsaturated fatty acids (but only in small amounts) fluidize the model membrane and increase its stability. Differences in the behavior between saturated and unsaturated fatty acids have been attributed to different shapes of molecules (rod-like stearic acid and coiled UFAs), resulting in either denser packing of cholesterol/DPPC monolayer or loosening its structure, respectively.

Studies on the influence of unsaturated fatty acid – linoleic acid - on model cholesterol-DPPC membrane was also performed by Makyła & Paluch (2009). In this paper, a model membrane that reflected the real proportion of these components in natural fibroblast cell membrane (cholesterol/DPPC = 0.36) and erythrocytes (cholesterol/DPPC = 0.82) was investigated upon the introduction of linoleic acid. Increased thermodynamic stability, evidenced by more negative ΔG^{exc} values) was observed for model fibroblast membrane, while the effect on erythrocytes membrane was significantly less pronounced.

4. Langmuir Monolayers Results *versus* the Biological Activity of PUFAs

The Langmuir monolayer technique, applied as an *in-vitro* platform to get insight into the biological function of unsaturated fatty acids, enabled to draw some important conclusions, drawn on the basis of the above-cited literature.

Firstly, it is important to underline that fatty acids have been proved to be surface active, which means that they have a tendency to gather at the interface (e.g., biomembrane) and incorporate into its structure, thus changing its physicochemical and mechanical properties, and – in consequence – modify its functioning.

Secondly, the geometry of fatty acid molecules, which is determined by the structure of the hydrocarbon chain (saturated; mono-unsaturated, poli-unsaturated; and *cis*- or *trans*- configuration of the unsaturated bonds) strongly affects their organization at the interface. As compared to saturated molecules, of rod-like structure, biologically important *cis*-unsaturated fatty acids show coiled conformation, and therefore occupy larger area at the surface as compared to saturated ones. Therefore, due to the incorporation of unsaturated fatty acids molecules, the biomembrane becomes more fluid. The coiled conformation of *cis*-unsaturated fatty acid molecules will thus increase molecular distances, which allow other molecules (for example cholesterol) to be easily incorporated into such a loose membrane structure. Indeed, it has been proved that the increased consumption of *cis*-unsaturated fatty acids increases the level of HDL, which captures excess cholesterol and transports it to the liver where it decomposes. Strong interactions between cholesterol and PUFAs, leading to the formation of stable complexes of unsaturated fatty acids and cholesterol, in which free cholesterol is bound, hinders in this way its deposition on artery walls. In such a way, PUFAs may impact in lowering cholesterol level, which results in beneficial effect, especially in cardiovascular health.

Experiments performed on fatty acids incorporated into 2-component (cholesterol-phospholipid) model membrane show significant differences in the behavior of saturated as compared to unsaturated fatty acids. The increased membrane rigidity upon introduction of saturated (stearic) acid makes the membrane less flexible and fragile. However – on the other hand – PUFAs were found to fluidize the membrane, thus "disturbing" the condensing effect of cholesterol on phospholipids. The fluidizing effect of PUFAs could also be harmful to some extent, i.e., it could result – in the extreme case - in membrane lysis. Thus, it is quite understandable that natural biomembrane – in order to maintain its fluidity on a desirable level – must regulate its phospholipid composition (proportion of saturated-to-unsaturated chains).

Another important conclusion is that the thermodynamic stability of model cholesterol-DPPC membrane was found to be stabilized by unsaturated fatty acids, but introduced only in low amounts. This can explain that free fatty acids are found in natural membranes in low proportion with respect to other lipids.

ACKNOWLEDGMENT

The authors are very grateful to Professor José Miñones Trillo *(University of Santiago de Compostela, Spain)* for his valuable remarks and fruitful discussion during the preparation of this manuscript.

REFERENCES

Adam, N. K., & Jessop, G. (1922). The Properties and Molecular Structure of Thin Films. Part III. Expanded Films. *Proceedings of the Royal*

Society A: Mathematical, Physical and Engineering Sciences, *101*(713), 516–531. https://doi.org/10.1098/rspa.1922.0063.

Adamson, A. W. (1967). *The Physical Chemistry of Surfaces* (2nd ed.). New York: Interscience,.

Aveyard, R., & Haydon, D. A. (1973). *An introduction to the principles of surface chemistry*. University Press.

Azzam, R. M. A., & Bashara, N. M. (1992). *Ellipsometry and polarized light*. North-Holland.

Bagatolli, L. A., Ipsen, J. H., Simonsen, A. C., & Mouritsen, O. G. (2010). An outlook on organization of lipids in membranes: Searching for a realistic connection with the organization of biological membranes. *Progress in Lipid Research*, *49*(4), 378–389. https://doi.org/10.1016/j.plipres.2010.05.001.

Brockman, H. (1999). Lipid monolayers: why use half a membrane to characterize protein-membrane interactions? *Current Opinion in Structural Biology*, *9*(4), 438–443. https://doi.org/10.1016/S0959-440X(99)80061-X.

Chan, Y.-H. M., & Boxer, S. G. (2007). Model membrane systems and their applications. *Current Opinion in Chemical Biology*, *11*(6), 581–587. https://doi.org/10.1016/j.cbpa.2007.09.020.

Crisp, D. J. (1949). *A two dimensional phase rule. I. Derivation of a two dimensional phase rule for plane interfaces*. London: Butterworths Scientific Publications.

Davies, J. T., & Rideal, E. K. (1963). Properties of Monolayers. In *Interfacial phenomena* (p. 276). New York: Academic Press.

de Mul, M. N. G., & Mann, J. A. (1998). Determination of the Thickness and Optical Properties of a Langmuir Film from the Domain Morphology by Brewster Angle Microscopy. *Langmuir*, *14*(9), 2455–2466. https://doi.org/10.1021/la9713154.

DeWolf, C., Leporatti, S., Kirsch, C., Klinger, R., & Brezesinski, G. (1999). Phase separation in phosphatidylinositol/phosphatidylcholine mixed monolayers. *Chemistry and Physics of Lipids*, *97*(2), 129–138. https://doi.org/10.1016/S0009-3084(98)00104-2.

Dynarowicz-Latka, P., Dhanabalan, A., & Oliveira, O. N. (2001). Modern physicochemical research on Langmuir monolayers. *Advances in Colloid and Interface Science*, *91*(2), 221–293. https://doi.org/10.1016/S0001-8686(99)00034-2.

Dynarowicz-Latka, P., & Hac-Wydro, K. (2014). Edelfosine in membrane environment - the Langmuir monolayer studies. *Anti-Cancer Agents in Medicinal Chemistry*, *14*(4), 499–508.

Dynarowicz-Latka, P., & Kita, K. (1999). Molecular interaction in mixed monolayers at the air/water interface. *Advances in Colloid and Interface Science*, *79*(1), 1–17. https://doi.org/10.1016/S0001-8686(98)00064-5.

Eeman, M., & Deleu, M. (2010). From biological membranes to biomimetic model membranes. In *Biotechnology, agronomy, society and environment* (pp. 1780–4507).

Feng, S. (1999). Interpretation of Mechanochemical Properties of Lipid Bilayer Vesicles from the Equation of State or Pressure–Area Measurement of the Monolayer at the Air–Water or Oil–Water Interface. *Langmuir*, *15*(4), 998–1010. https://doi.org/10.1021/la980144f.

Fidalgo Rodríguez, J. L. (2016). *Interacciones en monocapas de colesterol con ácidos grasos saturados e insaturados (omega-3, omega-6 y omega-9)*. [Interactions in monolayers of cholesterol with saturated and unsaturated fatty acids (omega-3, omega-6 and omega-9).] (PhD Thesis) Universidad de Santiago de Compostela. Retrieved from http://hdl.handle.net/10347/14627. Unpublished results.

Fidalgo Rodríguez, J. L., Dynarowicz-Latka, P., & Miñones Conde, J. (2017). Structure of unsaturated fatty acids in 2D system. *Colloids and Surfaces B: Biointerfaces*, *158*, 634–642. https://doi.org/10.1016/j.colsurfb.2017.07.016.

Gaines, G. L. (1966). *Insoluble monolayers at liquid-gas interfaces*. New York: Interscience Publishers.

Gong, K., Feng, S.-S., Go, M. L., & Soew, P. H. (2002). Effects of pH on the stability and compressibility of DPPC/cholesterol monolayers at the air-water interface. *Colloids and Surfaces A: Physicochemical and Engineering Aspects*, *207*(1–3), 113–125.

Gorter, E., & Grendel, F. (1925). On bimolecular layers of lipoids on the chromocytes of the blood. *The Journal of Experimental Medicine*, *41*(4), 439–43.

Hąc-Wydro, K., Jędrzejek, K., & Dynarowicz-Latka, P. (2009). Effect of saturation degree on the interactions between fatty acids and phosphatidylcholines in binary and ternary Langmuir monolayers. *Colloids and Surfaces B: Biointerfaces*, *72*(1), 101–111. https://doi.org/10.1016/j.colsurfb.2009.03.019.

Hac-Wydro, K., & Wydro, P. (2007). The influence of fatty acids on model cholesterol/phospholipid membranes. *Chemistry and Physics of Lipids*, *150*(1), 66–81. https://doi.org/10.1016/j.chemphyslip.2007.06.213.

Henon, S., & Meunier, J. (1991). Microscope at the Brewster angle: Direct observation of first-order phase transitions in monolayers. *Review of Scientific Instruments*, *62*(4), 936–939. https://doi.org/10.1063/1.1142032.

Holman, R. T. (1964). Nutritional and metabolic interrelationships between fatty acids. *Federation Proceedings*, *23*, 1062–7.

Hönig, D., & Möbius, D. (1991). Direct visualization of monolayers at the air-water interface by Brewster angle microscopy. *The Journal of Physical Chemistry*, *95*(12), 4590–4592. https://doi.org/10.1021/j100165a003.

Hughes, A. H. (1933). The behaviour of surface films of unsaturated compounds. *Journal of the Chemical Society*, 338. https://doi.org/10.1039/jr9330000338.

Iimura, K., Yamauchi, Y., Tsuchiya, Y., & Kato, T. (2001). 2-Dimensional Dendritic Growth of Condensed-Phase Domains in Spread Monolayers of Cis-Unsaturated Fatty-Acids. *Langmuir*, *17*(15), 4602–4609. https://doi.org/10.1021/la010236w.

Knobler, C. M. (1990). *Recent Developments in the Study of Monolayers at the Air-Water Interface* (pp. 397–449). John Wiley & Sons, Inc. https://doi.org/10.1002/9780470141267.ch7.

Langmuir, I. (1917). The constitution and fundamental properties of solids and liquids. II. Liquids. *Journal of the American Chemical Society*, *39*(9), 1848–1906. https://doi.org/10.1021/ja02254a006.

Linden, M., & Rosenholm, J. B. (1995). Influence of Multivalent Metal Ions on the Monolayer and Multilayer Properties of Some Unsaturated Fatty Acids. *Langmuir*, *11*(11), 4499–4504. https://doi.org/10.1021/la00011a053.

Maget-Dana, R. (1999). The monolayer technique: A potent tool for studying the interfacial properties of antimicrobial and membrane-lytic peptides and their interactions with lipid membranes. *Biochimica et Biophysica Acta - Biomembranes*, *1462*(1–2), 109–140. https://doi.org/10.1016/S0005-2736(99)00203-5.

Maillard, J. Y. (2002). Bacterial target sites for biocide action. *Journal of Applied Microbiology*, *92*(s1), 16S–27S. https://doi.org/10.1046/j.1365-2672.92.5s1.3.x.

Makyła, K., & Paluch, M. (2009). The linoleic acid influence on molecular interactions in the model of biological membrane. *Colloids and Surfaces B: Biointerfaces*, *71*(1), 59–66. https://doi.org/10.1016/j.colsurfb.2009.01.005.

Marsden, J., & Rideal, E. K. (1938). On monolayers of isomeric unsaturated compounds. *Journal of the Chemical Society*, 1163. https://doi.org/10.1039/jr9380001163.

Miñones Conde, J. (2001). *Estudio de las características estructurales y morfológicas de la anfotericina B, de fosfolípidos y de sus mezclas extendidos sobre la interfase aire-agua.* [Study of the structural and morphological characteristics of amphotericin B, of phospholipids and their extended mixtures on the air-water interface.] (PhD Thesis) Santiago de Compostela.

Moore, B., Knobler, C. M., Broseta, D., & Rondelez, F. (1986). Studies- of phase transitions in Langmuir monolayers by fluorescence microscopy. *Journal of the Chemical Society, Faraday Transactions 2*, *82*(10), 1753. https://doi.org/10.1039/f29868201753.

Nobre, T. M., Pavinatto, F. J., Caseli, L., Barros-Timmons, A., Dynarowicz-Latka, P., & Oliveira, O. N. (2015). Interactions of bioactive molecules & nanomaterials with Langmuir monolayers as cell membrane models. *Thin Solid Films*, *593*, 158–188. https://doi.org/10.1016/j.tsf.2015.09.047.

Overbeck, G. A., Hoenig, D., & Moebius, D. (1993). Visualization of first- and second-order phase transitions in eicosanol monolayers using Brewster angle microscopy. *Langmuir*, *9*(2), 555–560. https://doi.org/10.1021/la00026a032.

Overbeck, G. A., Hönig, D., Wolthaus, L., Gnade, M., & Möbius, D. (1994). Observation of bond orientational order in floating and transferred monolayers with Brewster angle microscopy. *Thin Solid Films*, *242*(1–2), 26–32. https://doi.org/10.1016/0040-6090(94)90496-0.

Pagano, R. E., & Gershfeld, N. L. (1972). A millidyne film balance for measuring intermolecular energies in lipid films. *Journal of Colloid and Interface Science*, *41*(2), 311–317. https://doi.org/10.1016/0021-9797(72)90116-6.

Peltonen, J. P., & Rosenholm, J. (1989). The influence of light on the properties of fatty acid-poly(3-octylthiophene) Langmuir-Blodgett films. *Thin Solid Films*, *179*(1–2), 543–547. https://doi.org/10.1016/0040-6090(89)90234-4.

Petty, M. C. (1996). *Langmuir-Blodgett films. An Introduction.* Cambridge: Cambridge University Press. https://doi.org/10.1017/CBO9780511622519.

Roberts, G. G. (Gareth G. (1990). *Langmuir-Blodgett films*. Plenum Press.

Schneider, V. L., Holman, R. T., & Burr, G. O. (1949). A Monolayer Study of the Isomerism of Unsaturated and Oxy Fatty Acids. *The Journal of Physical and Colloid Chemistry*, *53*(7), 1016–1029. https://doi.org/10.1021/j150472a004.

Seoane, R., Dynarowicz-Latka, P., Miñones Jr, J., & Rey-Gómez-Serranillos, I. (2001). Mixed Langmuir monolayers of cholesterol and `essential' fatty acids. *Colloid & Polymer Science*, *279*(6), 562–570. https://doi.org/10.1007/s003960000453.

Stefaniu, C., Brezesinski, G., & Möhwald, H. (2014). Langmuir monolayers as models to study processes at membrane surfaces. *Advances in Colloid and Interface Science*, *208*, 197–213. https://doi.org/10.1016/j.cis.2014.02.013.

Tomoaia-Cotişel, M., Zsako, J., Mocanu, A., Lupea, M., & Chifu, E. (1987). Insoluble mixed monolayers: III. The ionization characteristics of some fatty acids at the air/water interface. *Journal of Colloid and Interface Science*, *117*(2), 464–476. https://doi.org/10.1016/0021-9797(87)90407-3.

Vollhardt, D. (2007). Effect of Unsaturation in Fatty Acids on the Main Characteristics of Langmuir Monolayers. *The Journal of Physical Chemistry C*, *111*(18), 6805–6812. https://doi.org/10.1021/jp0704822.

Weidemann, G., Gehlert, U., & Vollhardt, D. (1995). Inner Structure of Condensed Phase Domains in Monolayers at the Air-Water Interface. *Langmuir*, *11*(3), 864–871. https://doi.org/10.1021/la00003a031.

Wu, S., & Huntsberger, J. R. (1969). Intermolecular interaction in some mixed polymer monolayers at the air water interface. *Journal of Colloid and Interface Science*, *29*(1), 138–147. https://doi.org/10.1016/0021-9797(69)90355-5.

In: Polyunsaturated Fatty Acids (PUFAs)
Editor: Angel Catala

ISBN: 978-1-53613-572-5

Chapter 4

LIPID PEROXIDATION AT THE CROSS ROADS OF CELL PROLIFERATION AND CANCER

Undurti N. Das*, MD
UND Life Sciences, WA, US

ABSTRACT

Polyunsaturated fatty acids (PUFAs) especially linoleic acid (LA, 18:2 n-6) and alpha-linolenic acid (ALA, 18:3 n-3) are widely distributed in various food items. Both LA and ALA are essential fatty acids (EFAs) that are present in various oils, leafy vegetables and other food items and are amenable to peroxidation during various processes including extraction, purification, and bleaching and cooking. Despite this, it is not clear what could be the biological consequences of lipid peroxides. In general, it is considered that lipid peroxides are harmful and are fraught with possible toxic effects on various tissues and cellular processes. Despite this belief, several studies revealed that lipid peroxides may have some beneficial actions on cell proliferation and cancer. It has been observed that formation of adequate amounts of lipid peroxides can inhibit cell proliferation especially those of cancer cells and smooth muscle cells of vascular tissue and thus may arrest cancer growth and

* Corresponding author E-mail: undurti@hotmail.com.

prevent atherosclerosis. On the other hand, formation of excess of lipid peroxides, especially in cancer cells, could lead to their apoptosis and thus, regression of cancer. These and other results suggest that one of the important factors that has a regulatory role in cell proliferation and cancer is lipid peroxides. Based on these evidences, it is suggested that efforts made to selectively generate lipid peroxides in cancer cells may lead to their growth inhibition and elimination of cancer whereas if adequate amounts of peroxides form in smooth muscle cells of vascular tissue can arrest atherosclerosis. These studies suggest that a delicate balance need to be maintained in the metabolism of EFAs and their long-chain metabolites such as arachidonic (AA), eicosapentaenoic (EPA) and docosahexaenoic (DHA) acids and their conversion to lipid peroxides and formation of their respective pro-and anti-inflammatory products, which have a regulatory role in cancer and atherosclerosis and other diseases.

INTRODUCTION

Lipid peroxidation (LP) process is believed to play a critical role in the oxidative damage to cell structures and lead to cell death. LP process involves formation and propagation of lipid radicals, the uptake of oxygen, a rearrangement of the double bonds in unsaturated lipids/fatty acids and the eventual destruction of membrane lipids. A variety of breakdown products including alcohols, ketones, alkanes, aldehydes and ethers are formed from LP both in plants and animals that can cause damage to proteins, DNA and lipids [1].

LP process can be defined as a chain reaction initiated by the hydrogen abstraction or addition of an oxygen radical, resulting in the oxidative damage of polyunsaturated fatty acids (PUFA), which can lead to an alteration in membrane fluidity and in the barrier functions of the membranes. LP and products formed by LP process such as hydroperoxides or their aldehyde derivatives have the ability to inhibit protein synthesis, alter macrophage functions/actions and modulate chemotactic signals and activities of a variety of enzymes [2]. A detailed discussion of the mechanisms of formation of LP due to the action of exogenous (such as radiation and anti-cancer drugs) and endogenous sources (such as from mitochondria, endoplasmic reticulum, peroxisomes) has been outlined previously [3]. In the present discussion, I will be highlighting only the biological significance of LP with specific focus on their clinical implications especially on cell proliferation and cancer.

PHYSIOLOGICAL SIGNIFICANCE OF FREE RADICALS AND LIPID PEROXIDES

The low and moderate concentrations of free radicals (ROS and RNS: reactive nitrogen species) may serve as potent signaling molecules in various types of cells including fibroblasts, endothelial cells, vascular smooth muscle cells, cardiac myocytes, etc. For instance, nitric oxide (NO) is an intercellular messenger for control of blood flow, thrombosis, and neural activity. NO also has a vital role in host defense, and has the capacity to kill several intracellular pathogens and tumor cells. Furthermore, free radicals are potent inducers of mitogenic response [4-6].

The secondary metabolites (aldehydes and ketones) formed during the formation of LP are likely to be toxic to cell. The purpose of intracellular metabolism of LP is to protect proteins from modification by aldehydic lipid peroxidation products. This process involves degradation and metabolism of LP leading to the formation of corresponding alcohol 1,4-dihydroxy-2-nonene (DHN), corresponding acid 4-hydroxy-2-nonenoic acid (HNA), and HNE-glutathione conjugate products. It is known that low and moderate levels of LP are probably needed for cell survival, while high levels can cause damage to proteins, induce autophagy and trigger senescence and thus, are toxic to cells.

In general, ROS are scavenged by antioxidant defenses of cell. This oxidative stress affects membrane lipids resulting in the release of PUFAs. These PUFAs especially: arachidonic acid (AA), eicosapentaenoic acid (EPA) and docosahexaenoic acid (DHA) are metabolized by COX-2 (cyclo-oxygenase-2) and LOX (lipoxygenase) enzymes (which are activated during inflammation) to form various prostaglandins (PGs), thromboxanes (TXs) and leukotrienes (LTs) that are known to have potent pro-inflammatory actions [7]. On the contrary, generation of excess of free radicals including ROS in tumor cells can overwhelm the cells' ability to neutralize them (free radicals) and thus, the overwhelmed antioxidant system fails and this ultimately results in their (tumor cells) apoptosis. In contrast, oxidative stress and generation of moderate amounts of lipid peroxides inside the tumor cells can promote cell migration, invasion, and metastasis due to further DNA damage leading several random mutations [8]. These evidences imply that the concentrations of ROS and the amounts of lipid peroxides generated may have profoundly different actions: there could occur enhancement of carcinognesis or apoptosis of cancer cells. In contrast to the actions of PGs, LTs and TXs (which have pro-inflammatory actions), lipoxins (LXs), resolvins, protectins and maresins derived from AA, EPA and DHA have potent anti-inflammatory actions and are critical for the resolution of inflammatory process and adequate wound healing. LXA4 (lipoxin A4) has been shown to prevent neutrophil infiltration, recruit and stimulate non-phagocytic monocytes, which phagocyte the dying neutrophils and facilitate resolution process. Macrophages also produce LXA4 to aid in clearing debris. LXA4 antagonizes the actions of LTB4 (a pro-inflammatory molecule), modulates ROS activated MAPK pathway, inhibits DNA-binding activity of NF-kB and thus brings about its anti-inflammatory action [9, 10].

Beneficial Actions of Free Radicals and LP

In general, it is believed that free radicals and LP have harmful actions, the final outcome of their Effects depends on the dose (concentration or level), timing and duration of exposure. For instance, at physiological levels, NO, a free radical, protects from malaria, cardiovascular diseases, acute inflammation, cancer and diabetes mellitus [3, 11, 12]. It is well documented that several growth factors {such as nerve growth factor (NGF), epidermal growth factor (EGF) and platelet derived growth factor (PDGF)}, cytokines (especially interleukin-6 and tumor necrosis factor-α), and other factors generate ROS by acting on their respective cell membrane receptors. This suggests that some of the actions of these endogenous molecules (growth factors, etc.,) is mediated by ROS. It is possible that such ROS generation leads to intracellular signaling resulting in enhancement of the corresponding transcription factors/signaling protein such as Rac1 that, in turn, activates many genes [3, 13-16].

ROS also seem to have a role in enhancing insulin sensitivity. ROS has a role in the phosphorylation of insulin receptor [17, 18] and hydrogen peroxide shows insulin like effect [19]. Insulin signaling increases production of ROS in the insulin targeted tissues [3, 19] implying that the signaling of insulin receptor is a co-regulatory process of ROS and insulin.

It has also been documented that at low or moderate concentrations, free radicals are necessary for normal immune response and host defense mechanisms of the body. Human leukocytes, monocytes, macrophages and T cells release free radicals to kill invading organisms, a process in which hydrogen peroxide seems to play a critical role [3, 20]. Physiological concentrations of superoxide and hydrogen peroxide enhance the production of interleukin-2 by stimulated T cells [20]. For instance, hydrogen peroxide in combination with anti-CD28 ligand enhances interleukin-2 production. Macrophages release different concentrations of prostaglandins, interleukin-6 and 12 based on the intracellular content of glutathione [21]. The ratio of T cells of type 1 and 2 is regulated by reductive and oxidative states of macrophages.

Similar to ROS, even LP shows some beneficial actions. LP plays a critical role in oxidative modification because reactive lipid-decomposition products trigger secondary modifications of proteins or nucleic acids. Of all the lipids, PUFAs such as DHA or EPA, have beneficial effects on human health, by serving as precursors of signaling molecules such as leukotrienes and/or prostaglandins, and also by increasing membrane fluidity. These PUFAs can be metabolized to form MDA under high oxidative stress conditions and thus, can show biological effects on the cells [3, 22].

When PUFAs are peroxidized by free radicals and enzymes (there could occur enzymatic and non-enzymatic peroxidation), reactive aldehydes, HNE, malondialdehyde (MDA), 4-oxo-2-nonenal (ONE), and glyoxal, are generated which can react with a protein can form a dihydropyridine type adduct that can be used as an indicator of oxidative stress. Of all the peroxidized products formed from PUFAs, MDA is the most commonly measured product of lipid peroxidation. When reactive oxygen and nitrogen species attack amino acids, lipids and carbohydrates, carbonyls groups are produced that can also be measured as an indicator of oxidative stress. Another measure that has been used to detect the extent of lipid peroxidation is F2-isoprostanes, isomers of prostaglandins produced by non-cyclooxygenase dependent peroxidation of arachidonic acid [3, 23].

At moderate concentrations, when the basal level of antioxidant enzymes are not sufficient to neutralize LP several transcription factors that are sensitive to stress such as nuclear factor erythroid 2-related factor 2 (Nrf2), activating protein-1 (AP-1), NF-κB, and peroxisome proliferator-activated receptors (PPAR) are altered in their expression. In addition, other stress response pathways such as mitogen-activated protein kinases (MAPK), EGFR/Akt pathways, and protein kinase C are also activated. Under physiological conditions of stress, Nrf2 is activated and translocated into the nucleus where it mediates the transcription of antioxidant/cytoprotective genes by binding to the antioxidant-response element (ARE) within DNA which has an important role in several pathological processes [24].

HARMFUL ACTIONS OF FREE RADICALS AND LIPID PEROXIDES

The harmful actions of free radicals and LP are well documented. In fact, most of the emphasis with regard to free radicals and LP has always been on their perceived harmful actions. The major diseases in which free radicals and LP have a role include cancer (by virtue of their ability to damage DNA and thus, activate oncogenes) [25, 26], diabetes mellitus (DM), especially in the pathogenesis of type 1 DM wherein IL-6, TNF-α, interferons, and other pro-inflammatory cytokines induce the release of free radicals and nitric oxide close to pancreatic βcells and destroy them to produce type 1 DM; and also have a role in the target organ damage seen in type 2 DM [27-29]; may underlie the occurrence of neurological disorders like Alzheimer's disease, Parkinson's disease, multiple sclerosis, amyotrophic lateral sclerosis (ALS), memory loss, and depression [30-32], and ageing [33].

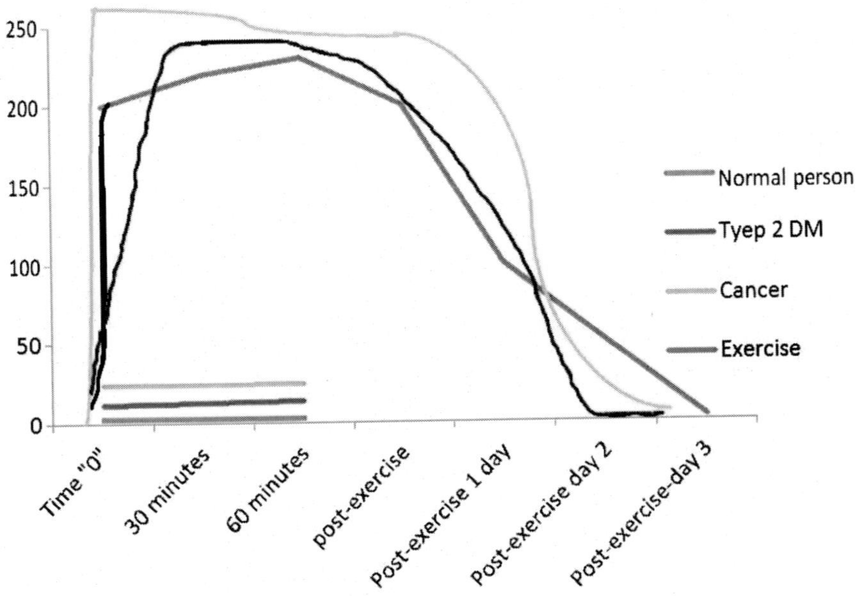

Refers to free radicals

Refers to plasma IL-6 levels

Refers to plasma TNF-α levels

In those with diabetes mellitus, plasma levels of IL-6, TNF-α and free radicals (measured as lipid peroxides) are persistently high compared to normal as shown in the figure.

Figure 1. Plasma IL-6 levels in normal, type 2 diabetes, patients with cancer and normal subjects with exercise (red line). In endurance exercise there will be an acute increase in free radicals, IL-6, and TNF-α followed by an abrupt and steady fall to normal after exercise.

In this context, it is interesting to note that exercise induces the release of significant amounts of free radicals, IL-6, and TNF-α suggesting that exercise is a state of free radical state. But, paradoxically exercise enhances the production of endogenous antioxidants such as SOD (superoxide dismutase) and thus, eventually exercise functions as an anti-inflammatory state. This suggests that there should occur certain amount of free radical generation to trigger the induction of a robust anti-oxidant defense in the body. It is noteworthy that exercise induces generation of large amounts of free radicals and substantial increase in IL-6 production (almost 200-300-fold increase) compared to control. But this increase in free radicals and IL-6 revert to normal within 24 hours (see Figure 1). On the other hand, in patients with cancer and diabetes mellitus, plasma levels of IL-6, TNF-α and free radicals (measured as LP or F2-isoprostanes) remain persistently elevated. Thus, it looks as though an abrupt increase and fall in the levels of IL-6, TNF-αand free radicals as is seen during exercise triggers the induction of a robust anti-oxidant system so that in the long run it produces beneficial actions as is seen following regular exercise. Thus, exercise can be considered as an anti-inflammatory trigger/status [34]. This ability of exercise to induce a robust anti-inflammatory state may also explain its ability to enhance telomere length [35, 36], and reduces cellular senescence [37]. These evidences suggest that possibly, a sudden sharp increase and an equally dramatic decrease in the circulating levels of free radicals, lipid peroxides, IL-6 and TNF-αas is seen during exercise suppresses aging process, whereas a relatively low or moderate but sustained increase in free radicals, lipid peroxides, and pro-inflammatory cytokines triggers and enhances the progression of aging as is seen in subjects with diabetes and cancer. Thus, free radicals and lipid peroxides have both beneficial and harmful actions depending on their plasma levels and the duration of their altered levels.

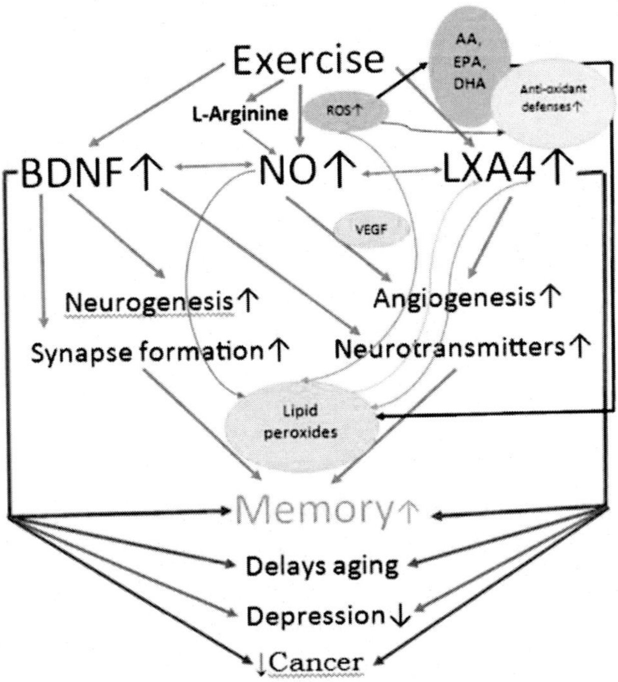

Figure 2. Scheme showing the actions of regular moderate exercise on memory, aging, depression and cancer incidence. Exercise enhances generation of free radical generation and formation of lipid peroxides, IL-6, TNF-α, BDNF, NO and LXA4. Increase in the generation of free radicals and lipid peroxides triggers upregulation of endogenous anti-oxidant defenses such as SOD catalase and glutathione. Exercise enhances the formation and release of endothelial NO, LXA4 formation (and possibly, resolvins, protectins and maresins) from their respective precursor PUFAs. LXA4 also enhances NO generation. BDNF enhances neurogenesis, synapse formation and thus improves memory. NO being a neurotransmitter, it plays a significant role on improving memory associated with exercise. NO and LXA4 improve angiogenesis to meet the much needed energy and oxygen demands of exercise. This is done by enhanced production of VEGF, a potent angiogenic molecule. It is posible that LXA4 may suppress the formation of excess of lipid peroxides whereas excess lipid peroxides suppress LXA4 formation. Similarly, superoxide anion inactivates NO and NO chelates superoxide anion. LXA4 and BDNF have cytoprotective actions and thus, posible cytotoxic actions of ROS generated during exercise is prevented by LXA4 and BDNF. Thus, a very delicate balance is maintained among free radicals, NO and LXA4 and their harmful and beneficial actions thus, ultimately all these molecules benefit the body. A better understanding of the molecular events occurring during exercise may have relevance to our understanding of pathophysiology of sepsis and septic shock and other critical illnesses such as ARDS (acute respiratory distress syndrome).

In addition to their relationship to free radicals cytokines and lipid peroxides, exercise also causes a substantial increase in the plasma and brain levels of BDNF (brain-derived neurotrophic factor), vascular endothelial growth factor (VEGF), lipoxin A4 (and possibly, that of resolvins, protectins and maresins), a potent anti-inflammatory metabolite of AA and nitric oxide. These four molecules (BDNF, VEGF, NO and LXA4) seem to enhance angiogénesis, neuronal regeneration and synapse formation and enhance neurotransmission that ultimately leads to improvement in memory, decreases the incidence of depression and aids in delaying the aging process (37, see Figure 2). It was also reported that lipid peroxides can trigger p53 pathway and thus, induce apoptosis of cells [3, 38]. This implies that lipid peroxides generated during exercise may induce the death of abnormal cells such as those that have DNA damage (like cancer cells) and this may explain as to why cancer incidence is low in those who perform regular exercise (see Figure 2).

PUFAs, Lipid Peroxides and Cell Proliferation and Its Implications to Cancer

PUFAs form an important constituents of cell membrane. PUFAs are also important constituents of the myelin sheath of the neuronal cells. PUFAs are the first in the line of attack by ROS to form peroxides. Yet, PUFAs can control ROS generation and ROS-induced inflammation. Arginine-induced nephrotoxicity that is due to free radical generation can be prevented by n3 PUFA in rats [39]. Oxidative stress is an important feature of sickle cell disease that can be controlled by EPA and DHA. Prenatal ethanol exposure induced neuronal damage due to reduced levels of glutathione can be prevented by supplementation of ω-3 fatty acids (EPA and DHA) that improved glutathione levels and reduced oxidative stress [40, 41].

In a series of studies, previously we and other showed that several PUFAs can selectively kill tumor cells with little effect on normal cells

[42-55]. This selective tumoricidal action has been ascribed to the ability of PUFAs (especially gamma-linolenic acid: GLA) reported to be by their ability to increase free radical generation and consequent lipid peroxidation process that seem to occur only in the tumor cells. Further studies showed that PUFAs control the expression of BCL2 gene and phosphorylation of BCL2, which indicates that in this specific instance PUFAs show antioxidant-like action in the tumor cells [52, 56]. Di-homo-gamma linolenic acid (DGLA), the precursor of AA, was noted to induce apoptosis of colon cancer cell by enhancing the expression of p53, a cancer suppressor gene and lipid peroxides formed as a result of free radical action of DGLA and other PUFAs enhanced the expression of caspases that, in turn, led to their apoptosis [57].

Our results showed that there is a direct relationship between the amount of lipid peroxides formed in the tumor cells and the degree of apoptosis of tumor cells [44, 49, 51-55]. These results suggest that the higher the amounts of lipid peroxides accumulate in the tumor cells, more or higher the number of tumor cells that undergo apoptosis. In other words, it implies that if tumor cells do not accumulate or fail to form adequate amount(s) of lipid peroxides they are unlikely to undergo apoptosis. This lack of accumulation of adequate amounts of lipid peroxides in the tumor cells could be due to deficiency of PUFAs (that need to be attacked by free radicals to generate lipid peroxides) and/or presence of excess of anti-oxidant vitamin E, an efficient anti-oxidant. Thus, the degree of apoptosis and the number of cells that can undergo apoptosis depends on the balance between the amount of PUFAs and vitamin E in the cell membranes. This is supported by the observation that tumor cells possess relatively higher amounts of vitamin E and are deficient in their PUFA content [1, 6, 8, 58]. In view of this, it is proposed that efforts made to enhance the PUFA content of tumor cells and/or generate more free radicals and lipid peroxides in the tumor cells would ultimately result in their apoptosis. This concept is supported by the recent evidences that showed that therapy-resistant tumor cells have low rates of lipid peroxidation or accumulate low amounts of lipid peroxides and/or have high levels of anti-oxidants such as glutathione and these resistant tumor cells become vulnerable to chemotherapy or radiation induced growth arrest or apoptosis when glutathione leves are reduced or the action of glutathione is inhibited [59, 60]. Since before the occurrence of apoptosis, tumor cells show growth arrest it is reasonable to suggest that PUFAs content, lipid peroxides and free radicals regulate mitosis. Thus, one of the factors that regulates cell proliferation is the concentrations of lipid peroxides in cells-the higher the lipid peroxide content in the cells the lower the cell proliferation index. Once the lipid peroxides in the cells reach optimal levels, it could induce apoptosis (see Figure 3).

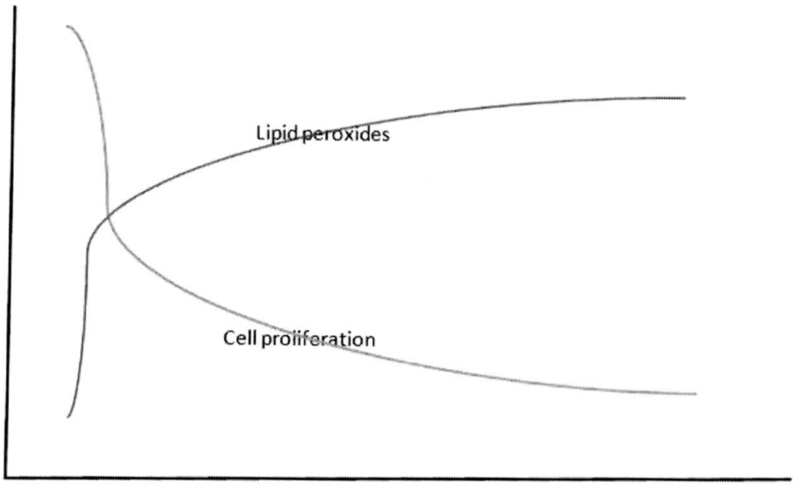

Figure 3. Scheme showing posible relationship between concentrations of lipid peroxides in the cells and the rate of cell proliferation. The higher the content of lipid peroxides in the cells, the lower the proliferation (mitosis) and vice versa. Note that no specific measurement values have been assigned to both X and Y axis. This so since, the formation of free radicals, lipid peroxides and the degree of inhibition of cell proliferation and apoptosis can vary from one cell type to the other, from tissue to tissue and all these are dynamic in nature and may vary from time to time. These may also depend on the state of the cell, amount of PUFAs, free radicals generated at a given point of time, lipid peroxides formed, anti-oxidant status of the cell and other factors and their half-life and potency and interaction(s) among them. Thus, the amount of lipid peroxides that induce apoptosis at one instance may not do so at another time. It is the question of balance among all the factors present a given point of time, the state of life cycle of the cell and all factors are dynamic in nature that can vary from time to time.

CONCLUSION

It is evident from the preceeding discussion that free radicals and lipid peroxides in moderate or physiological concentrations are needed for normal physiology. In instances wherein the levels of free radicals and lipid peroxides go beyond the physiological limits, then it leads to arrest of cell proliferation and apoptosis of cells (both normal and tumor cells) and consequently the initiation and perpetuation of inflammation and/or injury to various tissues, failure at resolution of inflammation and injury and hence, the persistance of diseases/disease process. In view of this, presence of an efficient antioxidant system is essential to protect the tissues from the oxidative toxicity/injury that could be achieved by regular moderate degree of exercise, consumption of adequate amounts of PUFAs and formation of anti-oxidant bioactive lipids such as lipoxins, resolvins, protectins and maresins. When the oxidative stress is in excess and antioxidant system fails, it could lead to tissue damage and various diseases.

Thus, free radicals and lipid peroxides have both physiological and pathological significance. In general, cells are endowed with a broad spectrum of responses to oxidative stress, depending on the stress type and level encountered. Oxidative stress that is beyond the capacity of the antioxidant defences induces oxidative damage, but low-level of stress is actually beneficial and enhances the defense capacity of the body.

Formation of lipid peroxides in response to generation of free radicals during various physiological and biochemical processes is a normal homeostatic process that may occur due to enzymatic or non-enzymatic and free-radical mediated reactions. The products and by-products of lipid peroxidation process have cytotoxic action, produce oxidative stress, oxidative damage and apoptosis. When this oxidative apoptosis occurs in tumor cells, it is beneficial and when this occurs in normal cells it is harmful. Thus, free radicals and lipid peroxides have both beneficial and harmful actions.

REFERENCES

[1] Dianzani, M. U., Barrera, G., Alvarez, S., Evelson, P., and Boveris, A. (2008). Pathology and physiology of lipid peroxidation and its carbonyl products. *Free radical pathophysiology*, 19-38.

[2] Mason, H. S., Fowlks, W. L., and Peterson, E. (1955). Oxygen transfer and electron transport by the phenolase complex1. *Journal of the American Chemical Society*, 77, 2914-2915.

[3] Naveen, G. V., Siresha, B., Das, U. N. (2017). Free radicals and lipid peroxides in health and disease. In: *Lipid peroxidation: Inhibition, Effects and Mechanisms*; Ed. Angel Catala, p. 99-124, Nova Science Publishers, NY.

[4] Pacher, P., Beckman, J. S., and Liaudet, L. (2007). Nitric oxide and peroxynitrite in health and disease. *Physiological reviews*, 87(1), 315-424.

[5] Genestra, M. (2007). Oxyl radicals, redox-sensitive signalling cascades and antioxidants. *Cellular signalling*, 19(9), 1807-1819.

[6] Halliwell, B. (1996). Antioxidants in human health and disease. *Annual review of nutrition*, 16(1), 33-50.

[7] Poorani, R., Bhatt, A. N., Dwarakanath, B. S., Das, U. N. (2016). COX-2, aspirin and metabolism of arachidonic, eicosapentaenoic and docosahexaenoic acids and their physiological and clinical significance. *Eur J Pharmacol* 785:116-132.

[8] Valko, M., Rhodes, C. J., Moncol, J., Izakovic, M. M., and Mazur, M. (2006). Free radicals, metals and antioxidants in oxidative stress-induced cancer. *Chemico-biological interactions*, 160(1), 1-40.

[9] Das, U. N. (2013). Lipoxins, resolvins, protectins, maresins and nitrolipids and their clinical implications with specific reference to cancer: Part I. *Clin Lipidol* 8, 437–463.

[10] Das, U. N. (2013). Lipoxins, resolvins, protectins, maresins and nitrolipids and their clinical implications with specific reference to diabetes mellitus and other diseases: Part II. *Clin Lipidol* 8: 465–480.

[11] Gaston, B., Drazen, J. M., Loscalzo, J., Stamler, J. S. (1994). The biology of nitrogen oxides in the airways. *American journal of respiratory and critical care medicine*, 149(2), 538-551.

[12] Änggård, E. (1994). Nitric oxide: mediator, murderer, and medicine. *The Lancet*, 343 (8907), 1199-1206.

[13] Bae, Y. S., Kang, S. W., Seo, M. S., Baines, I. C., Tekle, E., Chock, P. B., and Rhee, S. G. (1997). Epidermal growth factor (EGF)-induced generation of hydrogen peroxide Role in EGF receptor-mediated tyrosine phosphorylation. *Journal of Biological Chemistry*, 272(1), 217-221.

[14] Griendling, K. K., Sorescu, D., Lassègue, B., and Ushio-Fukai, M. (2000). Modulation of protein kinase activity and gene expression by reactive oxygen species and their role in vascular physiology and pathophysiology. *Arteriosclerosis, thrombosis, and vascular biology*, 20 (10), 2175-2183.

[15] Heeneman, S., Haendeler, J., Saito, Y., Ishida, M., and Berk, B. C. (2000). Angiotensin II induces transactivation of two different populations of the PDGFβ-receptor: key role for the adaptor protein Shc. *Journal of Biological Chemistry*.

[16] Bae, Y. S., Sung, J. Y., Kim, O. S., Kim, Y. J., Hur, K. C., Kazlauskas, A., and Rhee, S. G. (2000). Platelet-derived growth factor-induced H_2O_2 production requires the activation of phosphatidylinositol 3-kinase. *Journal of Biological Chemistry*, 275(14), 10527-10531.

[17] Hayes, G. R., and Lockwood, D. H. (1987). Role of insulin receptor phosphorylation in the insulinomimetic effects of hydrogen peroxide. *Proceedings of the National Academy of Sciences*, 84(22), 8115-8119.

[18] Heffetz, D., Rutter, W. J., and Zick, Y. (1992). The insulinomimetic agents H2O2 and vanadate stimulate tyrosine phosphorylation of potential target proteins for the insulin receptor kinase in intact cells. *Biochemical Journal*, 288(2), 631-635.

[19] May, J. M., and De Haen, C. (1979). Insulin-stimulated intracellular hydrogen peroxide production in rat epididymal fat cells. *Journal of Biological Chemistry*, 254(7), 2214-2220.

[20] Los, M., Dröge, W., Stricker, K., Baeuerle, P. A., and Schulze-Osthoff, K. (1995). Hydrogen peroxide as a potent activator of T lymphocyte functions. *European journal of immunology*, 25(1), 159-165.

[21] Hamuro, J., Murata, Y., Suzuki, M., Takatsuki, F., and Suga, T. (1999). The triggering and healing of tumor stromal inflammatory reactions regulated by oxidative and reductive macrophages. *Gann monograph on cancer research*, 48, 153-164.

[22] Murphy, M. E., and Sies, H. (1991). Reversible conversion of nitroxyl anion to nitric oxide by superoxide dismutase. *Proceedings of the National Academy of Sciences*, 88 (23), 10860-10864.

[23] Nair, U.; Barstsch, H. and Nair, J. (2007) Lipid peroxidation-induced DNA damage in cancer prone inflammatory diseases: a review of published adduct types and levels in humans. *Free Radical in Biology and Medicine.* 43, 1109-1120.

[24] Gan, L., and Johnson, J. A. (2014). Oxidative damage and the Nrf2-ARE pathway in neurodegenerative diseases. *Biochimica et Biophysica Acta (BBA)-Molecular Basis of Disease*, 842(8), 1208-1218.

[25] Suh, Y. A., Arnold, R. S., Lassegue, B., Shi, J., Xu, X., Sorescu, D., and Lambeth, J. D. (1999). Cell transformation by the superoxide-generating oxidase Mox1. *Nature*, 401 (6748), 79-82.

[26] Hack, V., Schmid, D., Breitkreutz, R., Stahl-Henning, C., Drings, P., Kinscherf, R., and Dröge, W. (1997). Cystine levels, cystine flux, and protein catabolism in cancer cachexia, HIV/SIV infection, and senescence. *The FASEB journal*, 11(1), 84-92.

[27] Baynes, J. W. (1991). Role of oxidative stress in development of complications in diabetes. *Diabetes*, 40(4), 405-412.

[28] Nishikawa, T., Edelstein, D., Du, X. L., Yamagishi, S. I., Matsumura, T., Kaneda, Y., et al. (2000). Normalizing mitochondrial superoxide production blocks three pathways of hyperglycaemic damage. *Nature*, 404(6779), 787-790.

[29] Das, U. N. (2017) Is there a role for bioactive lipids in the pathobiology of diabetes mellitus. *Front Endocrinol (Lausanne),* 8, 182.

[30] Multhaup, G., Ruppert, T., Schlicksupp, A., Hesse, L., Beher, D., Masters, C. L., and Beyreuther, K. (1997). Reactive oxygen species and Alzheimer's disease. *Biochemical Pharmacology*, 54(5), 533-539.

[31] Pasinelli, P., Houseweart, M. K., Brown, R. H., and Cleveland, D. W. (2000). Caspase-1 and-3 are sequentially activated in motor neuron death in Cu, Zn superoxide dismutase-mediated familial amyotrophic lateral sclerosis. *Proceedings of the National Academy of Sciences*, 97 (25), 13901-13906.

[32] Maurice, M. M., Nakamura, H., Van Der Voort, E. A., van Vliet, A. I., Staal, F. J., Tak, P. P., et al. (1997). Evidence for the role of an altered redox state in hyporesponsiveness of synovial T cells in rheumatoid arthritis. *The Journal of Immunology*, 158(3), 1458-1465.

[33] Beckman, K. B., and Ames, B. N. (1998). The free radical theory of aging matures. *Physiological Reviews*, 78(2), 547-581.

[34] Das UN. (2004). Anti-inflammatory nature of exercise. *Nutrition,* 20, 323-326.

[35] Diman, A., Boros, J., Poulain, F., Rodriguez, J., Purnelle, M., Episkopou, H. (2016). Nuclear respiratory factor 1 and endurance exercise promote human telomere transcription. *Sci. Adv.,* 2, e1600031.

[36] Saßenroth, D., Meyer, A., Salewsky, B., Kroh, M., Norman, K., Steinhagen-Thiessen, E., Demuth, J. (2015). Sports and Exercise at Different Ages and Leukocyte Telomere Length in Later Life – Data from the Berlin Aging Study II (BASE-II). *PLoS ONE* 10, e0142131. doi: 10.1371/journal.pone.0142131.

[37] Schafer, M. J., White, T. A., Evans, G., Tonne, J. M., Verzosa, G. C., Stout, M. B., Mazula, D. L., et al. (2016). Exercise Prevents Diet-induced Cellular Senescence in Adipose Tissue. *Diabetes.* 65, 1606-1615.

[38] Shibata, T., Iio, K., Kawai, Y., Shibata, N., Kawaguchi, M., Toi, S., et al. (2006). Identification of a Lipid Peroxidation Product as a Potential Trigger of the p53 Pathway. *Journal of Biological Chemistry.* 281, 1196–1204.

[39] Khan, M. W., Priyamvada, S., Khan, S. A., Khan, S., Naqshbandi, A., and Yusufi, A. N. K. (2012). Protective effect of ω-3 polyunsaturated fatty acids on L-arginine-induced nephrotoxicity and oxidative damage in rat kidney. *Human and experimental toxicology*, 31(10), 1022-1034.

[40] Daak, A. A., Ghebremeskel, K., Mariniello, K., Attallah, B., Clough, P., and Elbashir, M. I. (2013). Docosahexaenoic and eicosapentaenoic acid supplementation does not exacerbate oxidative stress or intravascular haemolysis in homozygous sickle cell patients. *Prostaglandins, Leukotrienes and Essential Fatty Acids (PLEFA)*, 89(5), 305-311.

[41] Patten, A. R., Brocardo, P. S., and Christie, B. R. (2013). Omega-3 supplementation can restore glutathione levels and prevent oxidative damage caused by prenatal ethanol exposure. *The Journal of nutritional biochemistry*, 24, 760-769.

[42] Begin, M. E., Ells, G., Das, U. N., Horrobin, D. F. (1986). Differential killing of human carcinoma cells supplemented with n-3 and n-6 polyunsaturated fatty acids. *J Natl Cancer Inst* 77, 1053-1062.

[43] Das, U. N. (2006). Effect of γ- Linolenic acid on the transcriptional activity of the HER-2neu (erbB-2) oncogene. *J Natl Cancer Inst*, 98, 2006.

[44] Das, U. N. (1991). Tumoricidal action of cis-unsaturated fatty acids and its relationship to free radicals and lipid peroxidation. *Cancer Lett,* 56, 235-243.

[45] Ge, H., Kong, X., Shi, L., Hou, L., Liu, Z., Li, P. (2009). Gamma-linolenic acid induces apoptosis and lipid peroxidation in human chronic myelogenous leukemia K562cells. *Cell Biol Int,* 33, 402-410.

[46] Begin, M. E., Das, U. N., Ells, G. (1986). Cytotoxic effects of essential fatty acids (EFA) in mixed cultures of normal and malignant human cells. *Prog Lipid Res.* 25, 573-576.

[47] Madhavi, N., Das, U. N. (2011). Effect of polyunsaturated fatty acids on drug sensitive and resistant tumor cells in vitro. *Lipids in health and disease*, 10, 159.

[48] Gillis, R. C., Daley, B. J., Enderson, B. L., Karlstad, M. D. (2002). Eicosapentaenoic acid and gamma-linolenic acid induce apoptosis in HL-60 cells. *J Surg Res* 107, 145-153.

[49] Madhavi, N., Das, U. N. (1994). Effect of n-6 and n-3 fatty acids on the survival of vincristine sensitive and resistant human cervical carcinoma cells in vitro. *Cancer Lett,* 84, 31-41.

[50] Trombetta, A., Maggiora, M., Martinasso, G., Cotogni, P., Canuto, R. A., Muzio, G. (2007). Arachidonic and docosahexaenoic acids reduce the growth of A549 human lung-tumor cells increasing lipid peroxidation and PPARs. *Chem Biol Interact,* 165, 239-250.

[51] Das, U. N., Begin, M.E., Ells, G., Horrobin, D. F. (1987). Uptake and distribution of cis-unsaturated fatty acids and their effect on free radical generation in normal and tumor cells in vitro. *Free Rad Biol Med,* 3, 9-16.

[52] Das, U. N., Begin, M. E., Ells, G., Horrobin, D. F. (1987). Polyunsaturated fatty acids augment free radical generation in tumor cells in vitro. *Biochem Biophys Res Commun.* 145, 15-20.

[53] Sagar, P. S., Das, U, N. (1995). Cytotoxic action of cis-unsaturated fatty acids on human cervical carcinoma (HeLa) cells in vitro. *Prostaglandins Leukot Essent Fatty Acids* 53, 287-299.

[54] Sagar, P. S., Das, U, N., Koratkar, R., Ramesh, G., Padma, M., Kumar, G. S. (1992). Cytotoxic action of cis-unsaturated fatty acids on human cervical carcinoma (HeLa) cells: relationship to free radicals and lipid peroxidation and its modulation by calmodulin antagonistas. *Cancer Lett,* 63, 189-198.

[55] Das, U. N. (1999). Essential fatty acids, lipid peroxidation and apoptosis. *Prostaglandins, Leukotrienes and Essential Fatty Acids.* 61, 157-163.

[56] Xu, Y., Qi, J., Yang, X., Wu, E., and Qian, S. Y. (2014). Free radical derivatives formed from cyclooxygenase-catalyzed dihomo-γ-linolenic acid peroxidation can attenuate colon cancer cell growth and enhance 5-fluorouracil's cytotoxicity. *Redox biology*, 2, 610-618.

[57] Das, U. N., Madhavi, N. (2011). Effect of polyunsaturated fatty acids on drug-sensitive and resistant tumor cells *in vitro*. *Lipids Health Dis.* 10, 159.

[58] Das, U. N. (2011). Essential fatty acids enhance free radical generation and lipid peroxidation to induce apoptosis of tumor cells. *Clin Lipidol.* 6, 463–489.

[59] Viswanathan, V. S., Ryan, M. J., Dhruv, H. D., Gill, S., Eichhoff, O. M., Seashore-Ludlow, B., et al. (2017). Dependency of a therapy-resistant state of cancer on a lipid peroxide pathway. *Nature.* 547, 453-457.

[60] Hangauer, M., Viswanathan, V. S., Ryan, M. J., Dhruv, B., Eaton, J. E., Matov, A., et al. (2017). Drug-tolerant persister cancer cells are vulnerable to GPX-4 inhibition. *Nature.* 551, 247-250.

In: Polyunsaturated Fatty Acids (PUFAs) ISBN: 978-1-53613-572-5
Editor: Angel Catala

Chapter 5

ADVANCEMENT IN THE KNOWLEDGE OF POLYUNSATURATED FATTY ACIDS, FROM DISCOVERY TO NOW

Angel Catalá*

Instituto de Investigaciones Fisicoquímicas Teóricas y Aplicadas, (INIFTA-CCT La Plata-CONICET), Facultad de Ciencias Exactas, Universidad Nacional de La Plata, Argentina
The author is Member of Consejo Nacional de Investigaciones Científicas y Técnicas (CONICET), Argentina

ABSTRACT

Polyunsaturated fatty acids (PUFAs) are fatty acids that contain more than one double bond in their backbone.

The most important polyunsaturated fatty acids can be grouped in two families: n-3 and n-6 fatty acids are a group of fatty acids, which have the first unsaturated bond in the third and sixth position respectively from the omega carbon.

---

* Corresponding Email: catala@inifta.unlp.edu.ar.

The most important polyunsaturated fatty acids can be described as follow:

α-Linolenic acid C18:3n-3 is a polyunsaturated fatty acid (PUFA). It is a member of the group of essential fatty acids called n-3 fatty acids. α-Linolenic acid (18:3n-3) is the predominant plant-derived nutritional n-3 PUFA and is a precursor for docosahexaenoic acid 22:6n-3 (DHA).

Linoleic acid (18:2n-6) is the predominant plant-derived nutritional n-6 PUFA and is a precursor for arachidonic acid (20:4n-6) and eicosanoids.

The first evidences for the existence of polyunsaturated fatty acids containing a very high iodine number was first described in the late 1800s in studies of fish oils by (Bull; 1989) and (Tsujimoto; 1906).

The first isolation and characterization of a tetraunsaturated 20-carbon fatty acid was reported by a young investigator at the Lister Institute, Percival Hartley, who was working on his D.Sc. degree from the University of London (Dale; 1957).

In 1929, George Oswald Burr and his wife Mildred, published a paper (Burr and Burr, 1929) which discovered that elimination of fat from the diet of animals induced a deficiency illness, and their afterward papers showed that this illness could be prevented or cured by the addition of linoleic acid in the diet (Burr and Burr, 1930) and (Burr, 1942). Thus, they proved convincingly that linoleic acid was an essential fatty acid.

In this chapter I would like to review some basic concepts of the chemistry and biochemistry of polyunsaturated fatty acids, and then I will review some selective parts of the research I was involved that range from the early sixties up to now.

Keywords: polyunsaturated fatty acids (PUFAs), α-Linolenic acid, n-3 PUFA, docosahexaenoic acid, Linoleic acid, n-6 PUFA, Eicosapentaenoic acid, arachidonic acid

1. Introduction

My first experience with polyunsaturated fatty acids started in 1964 when I was accepted as "Research Assistant" without salary at the Cátedra de Bioquímica, Instituto de Fisiología, Facultad de Ciencias Médicas, Universidad Nacional de La Plata, Argentina. This was before the era of molecular biology and the limitations in biochemical science were organic and analytical chemistry. My tutor was Professor Rodolfo R. Brenner, a previous pupil of Professor Pedro Cattaneo who worked for many years with vegetable oils. R. Brenner a Researcher Emeritus of CONICET and Professor Titular Emeritus of the National University of La Plata, created in 1965, together with Drs. Federico Leloir (Nobel prize of Chemistry, 1970) and Andres Stoppani (Tutor of Cesar Milstein-Novel prize in Physiology or Medicine 1984), The Society Argentina of Biochemistry Research. Seventeen years later, Brenner established the Institute for Biochemical Research of La Plata whose foundations date back to 1956 and serves as a double dependency CONICET/UNLP since 1982. Its main objective is the investigation of biological, biochemical, and biophysical processes related to lipid metabolism in various organisms.

Polyunsaturated fatty acids have followed me during my whole scientific career and I have published a number of studies concerned with different aspects of them such as chemical synthesis, mechanism of enzymatic formation, metabolism, transport, physical, chemical, and catalytic properties of a reconstructed desaturase system in liposomes, lipid peroxidation and its biological implications, and quantitative methods for its analysis.

In this chapter, I would like to review some basic concepts of the chemistry and biochemistry of n-3 and n-6 PUFAs, and then I will review some selective parts of the research I was involved in that ranges from the early sixties up to now.

2. Polyunsaturated Fatty Acids Can Be Classified in Various Groups by Their Chemical Structure

Polyunsaturated fatty acids (PUFAs) are fatty acids that contain more than one double bond in their backbone. The most important polyunsaturated fatty acids can be grouped in two families: n-3 and n-6 fatty acids are a group of fatty acids, which have the first unsaturated bond in the third and sixth position respectively from the omega carbon.

Linoleic Acid C18:2n-6

Linoleic acid is a doubly unsaturated fatty acid, also known as an n-6 fatty acid, occurring widely in plant glycosides. In this particular polyunsaturated fatty acid (PUFA), the first double bond is located between the sixth and seventh carbon atom from the methyl end of the fatty acid (n-6). Linoleic acid is an essential fatty acid in human nutrition because it cannot be synthesized by humans. It is used in the biosynthesis of prostaglandins (via arachidonic acid) and cell membranes. (From Stedman, 26th ed).

α-Linolenic Acid C18:3n-3

α-Linolenic acid was first isolated by Rollett (Rollett; 1909) as cited in (McCutcheon's; 1942) and referred to in Green and Hilditch's 1930s survey. (Green and Hilditch; 1935). It was first artificially synthesized in 1995 from C6 homologating agents. A Wittig reaction of the phosphonium salt of ((Z-Z)-nona-3,6-dien-1-yl)triphenylphosphonium bromide with methyl 9-oxononanoate, followed by saponification, completed the synthesis.

Alpha-linolenic acid (ALA) is a polyunsaturated fatty acid (PUFA). It is a member of the group of essential fatty acids called n-3 fatty acids. Alpha-linolenic acid, in particular, is not synthesized by mammals and therefore is an essential dietary requirement for all mammals. Certain nuts (English walnuts) and vegetable oils (canola, soybean, flaxseed/linseed, olive) are particularly rich in alpha-linolenic acid (ALA). N-3 fatty acids get their name based on the location of one of their first double bond. In all n-3 fatty acids, the first double bond is located between the third and fourth carbon atom counting from the methyl end of the fatty acid (n-3). Although humans and other mammals can synthesize saturated and some monounsaturated fatty acids from carbon groups in carbohydrates and proteins, they lack the enzymes necessary to insert a cis double bond at the n-6 or the n-3 position of a fatty acid. N-3 fatty acids like -linolenic acids are important structural components of cell membranes. When incorporated into phospholipids, they affect cell membrane properties such as fluidity, flexibility, permeability and the activity of membrane bound enzymes. N-3 fatty acids can modulate the expression of a number of genes, including those involved with fatty acid metabolism and inflammation. Alpha linolenic acid and other omega 3 fatty acids may regulate gene expression by interacting with specific transcription factors, including peroxisome proliferator activated receptors (PPARs) and liver X receptors (LXRs).

Arachidonic Acid C20:4n-6

Arachidonic Acid is an unsaturated, essential fatty acid. It is found in animal and human fat as well as in the liver, brain, and glandular organs, and is a constituent of animal phosphatides. It is formed by the synthesis from dietary alpha-linoleic acid and is a precursor in the biosynthesis of prostaglandins, thromboxanes, and leukotrienes.

Discovery of Arachidonic Acid

The existence of polyunsaturated fatty acids containing a very high iodine number was first described in the late 1800s in studies of fish oils by (Bull; 1989) and (Tsujimoto; 1906). However, these investigators did not characterize in chemical detail these polyunsaturated fatty acids except to state that they were of the series $CnH_2n\text{-}8O_2$, in which arachidonate belongs. However, measurement of the iodine value (grams of iodine that reacts with 100 g of fatty acid) of somewhat purified polyunsaturated fatty acids was found to be above 300 for fatty acid components in herring oil (Bull; 1989).

The first isolation and characterization of a tetraunsaturated 20-carbon fatty acid was reported by a young investigator at the Lister Institute, Percival Hartley, who was working on his D.Sc. degree from the University of London (Dale; 1957).

Eicosapentaenoic Acid C20:5n-3

Important polyunsaturated fatty acid found in fish oils. Eicosapentaenoic acid serves as the precursor for the prostaglandin-3 and thromboxane-3 families. A diet rich in eicosapentaenoic acid lowers serum lipid concentration, reduces incidence of cardiovascular disorders, prevents platelet aggregation, and inhibits arachidonic acid conversion into the thromboxane-2 and prostaglandin-2 families.

Docosahexaenoic Acid C22:6n-3

Docosahexaenoic acid (DHA) is an n-3 essential fatty acid. Chemically, DHA is a carboxylic acid with a 22-carbon chain and six cis double bonds with the first double bond is located at the third carbon from the omega end. DHA is most often found in fish oil. It is a major fatty acid in sperm and brain phospholipids, especially in the retina. Dietary DHA can reduce the level of blood triglycerides in humans, which may reduce the risk of heart disease.

3. The Discovery of Essential Fatty Acids

In the years 1927 and 1929, two discoveries were made almost at the same time: vitamin E (Evans et al., 1927) and the essential unsaturated fatty acids (Burr and Burr, 1929) Vitamin E was introduced with great display and was a continuous preferred of the media, but there is little doubt today that the polyunsaturated fatty acids are a more significant dietary component than is vitamin E.

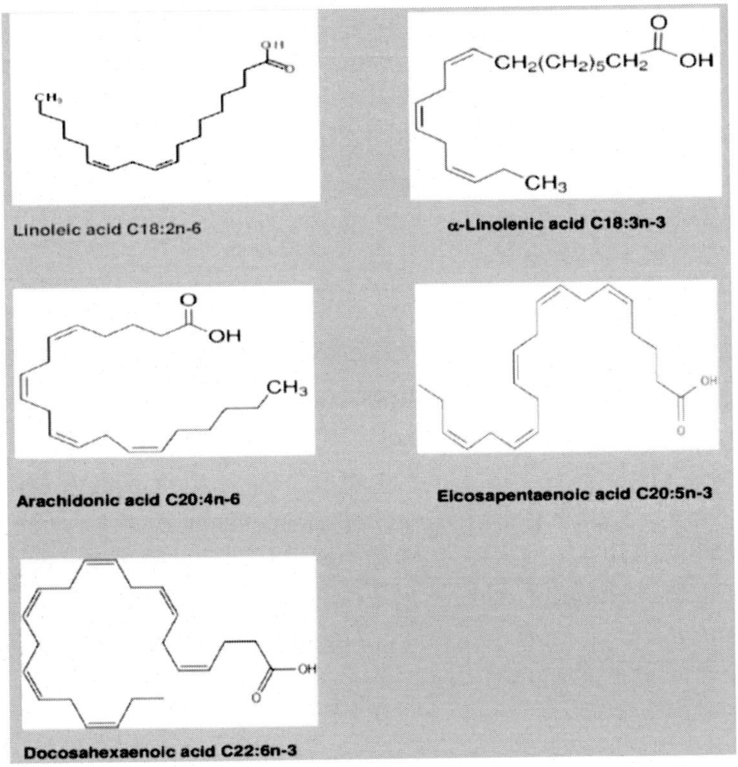

Figure 1. Simplified *chemical structure* of polyunsaturated fatty acids.

Linoleic acid (18:2n-6) is the predominant plant-derived nutritional n-6 PUFA and is a precursor for arachidonic acid (20:4n-6) and eicosanoids. α-Linolenic acid (18:3n-3) is the predominant plant-derived nutritional n-3 PUFA and is a precursor for docosahexaenoic acid 22:6n-3 (DHA).

During the last four decades the interest in polyunsaturated fatty acids has augmented manifolds and the number of published studies is rising each year. The current impetus for this interest has been mainly the observation that polyunsaturated fatty acids (PUFAs) and their metabolites have a diversity of physiological roles including: energy provision, membrane structure, cell signaling and regulation of gene expression. In addition the observation that PUFAs are targets of lipid peroxidation opens a new important area of investigation. Lipid peroxidation is one of the major outcomes of free radical-mediated injury to tissue. Peroxidation of fatty acyl groups occurs mostly in membrane phospholipids. Peroxidation of lipids can greatly alter the physicochemical properties of membrane lipid bilayers, resulting in severe cellular dysfunction. In addition, a variety of lipid byproducts are produced as a consequence of lipid peroxidation, some of which can exert adverse and/or beneficial biological effects.

4. Chemistry and Biochemistry of n-3 and n-6 PUFAs: Some Basic Concepts

Fatty acids are constituted by hydrocarbon chains with a carboxyl group at one end and a methyl group at the opposite end (n or ω end). The unsaturated fatty acids hold in their chain one or more double bonds not saturated with hydrogen. PUFAs contain more than one of these double bonds, and the n-3 or n-6 designation indicates that the first double bond on the hydrocarbon chain is located at the third and sixth carbons from the n (or ω) end, respectively.

Commonly denoted as "omega 3 or 6," the n-3 and n-6 polyunsaturated fatty acids (n-3 and n-6 PUFAs) are constituents of our diet found both in vegetables and in animal tissues. However, they are predominantly concentrated in fish, which is commonly considered the best source for these fatty acids (especially the long-chain n-3 PUFAs) (Marszalek and Lodish, 2005), whereas n-6 PUFAs belong to another significant family of dietary PUFAs and are mainly found in vegetable oils.

Among the components of these two classes of PUFAs (n-3 and n-6), α-linolenic acid (α-LNA, 18:3 n-3) and linoleic acid (LA, 18:2 n-6) are considered essential for our diet. This is due to the fact that, differently from vegetable cells, mammalian cells do not have the desaturases able to specifically put the double bonds in n-3 and n-6 positions on the fatty acid carbon chain along the synthetic pathway (Knutzon et al., 1998). From that point onward, our cells are instead supplied with all the enzymes required to create longer and more unsaturated n-3 or n-6 PUFAs from either α-LNA or LA (Δ6-desaturase and Δ5-desaturase and elongase).

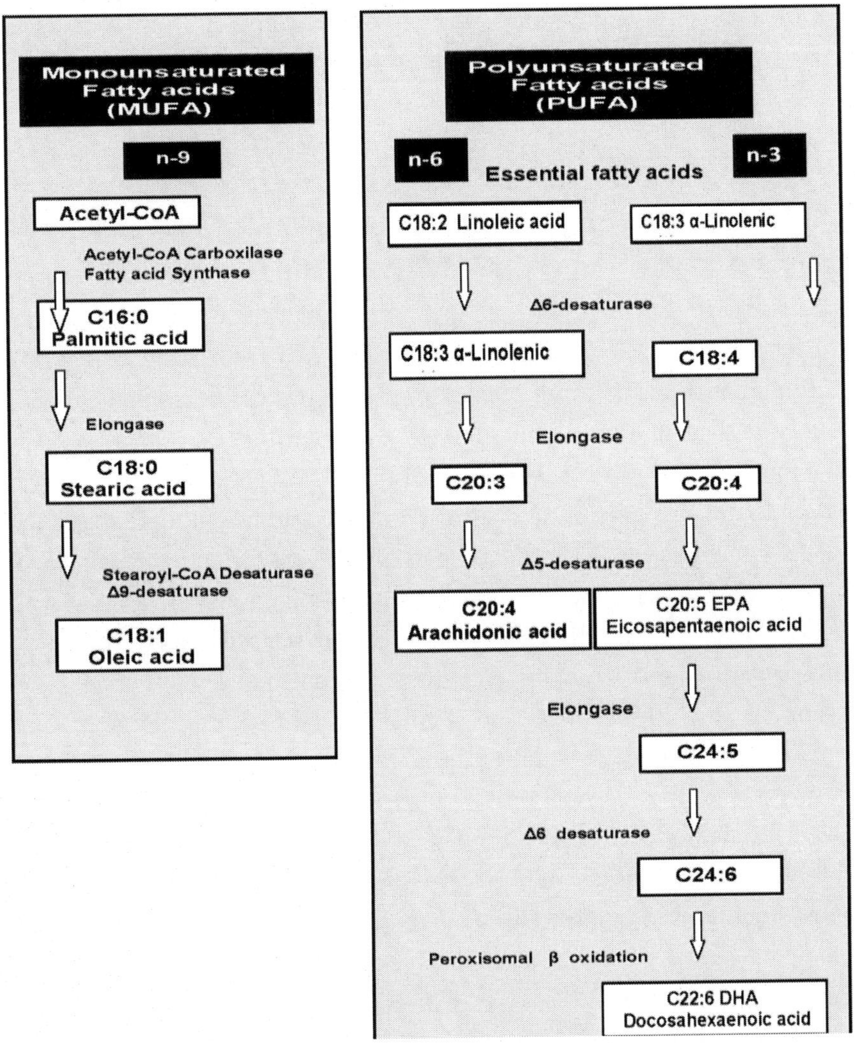

Figure 2. Synthesis-of-unsaturated-fatty-acids-in-mammals.

Two completely different synthetic pathways exist for the construction of the long-chain n-3 and n-6 PUFAs (Figure 2), and a component of one class is not exchangeable into a PUFA of the other class.

Through a number of synthetic steps, the n-6 PUFA arachidonic acid (AA, 20:4 n-6) is produced, which is highly founded in animal membranes and is the precursor of a lot of bioactive compounds normally included under the name of eicosanoids, an expression derived from the Greek eicosa (twenty), signifying the 20 carbons of the AA chain. Among the eicosanoids, there are prostaglandins, thromboxanes, and leukotrienes.

Lipid Peroxidation of Membrane Phospholipids Generates Hydroxy-Alkenals and Oxidized Phospholipids Active in Physiological and/or Pathological Conditions

Membrane phospholipids containing polyunsaturated fatty acids are particularly susceptible to oxidation and can contribute in chain reactions that amplify damage to bio molecules. Lipid peroxidation often occurs in response to oxidative stress, and a great diversity of phospholipid oxidation products and aldehydes is formed when lipid hydroperoxides break down in biological systems. Bioactivities of these phospholipids on vascular wall cells, leukocytes, and platelets have been described. Some of these aldehydes are highly reactive and may be considered as second toxic messengers which disseminate and augment initial free radical events. The aldehydes most intensively studied up to now are 4-hydroxy-2-nonenal and 4-hydroxy-2-hexenal. 4-Hydroxy-2-nonenal (HNE) is known to be the main aldehyde formed during lipid peroxidation of n-6 polyunsaturated fatty acids, such as linoleic acid C18:2 n-6 and arachidonic acid C20:4 n-6 whereas lipid peroxidation of n-3 polyunsaturated fatty acids such as α-linolenic acid C18:3 n-3 and docosahexaenoic acid C22:6 n-3 generates a closely related compound, 4-hydroxy-2-hexenal (HHE). During lipid peroxidation, biomolecules such as proteins or amino lipids can be covalently modified by these lipid decomposition products, which damage membrane structure modifying its physical properties. In addition this review provide a synopsis of identified effects of hydroxy-alkenals and oxidized phospholipids on cell signaling, from their intracellular production, to their action as intracellular messenger, up to their influence on transcription factors and gene expression (Catalá, 2009).

The Lipid Peroxidation Process Produces Oxidized Phospholipids that Acquire New Biological Activities Not Characteristic of Their Unoxidized Precursors

Membranes form selective barriers that separate, communicate and define cells and their internal organelles, and they also receive and propagate important signals that control cellular behavior. Biomembranes contains different phospholipid classes (head group heterogeneity), subclasses (acyl, alkyl chains) and species (chain length and unsaturation degree). PC is the major phospholipid in all mammalian cells (40–50%) and thus, most oxidized phospholipids detected in mammalian tissues have the choline moiety. However, recently oxidized PE has been found in the retina, a tissue that contains very high quantities of ethanolamine lipids (Gugiu et al., 2006). enriched in docosahexaenoic acid (Guajardo et al., 2006). In addition, there are also reports providing evidence for the presence of oxidized PS in the surface of apoptotic cells (Matsura et al., 2005). Furthermore, oxidized phospholipids have been demonstrated to act as signals in monocyte activation, programmed cell death, and phagocytotic clearance of apoptotic cells (Cole et al., 2003; Greenberg et al., 2006; Maskrey et al., 2007; Tyurina et al., 2004; Walton et al., 2003).

In eukaryotic phospholipids, the sn-1 position is either linked to an acyl residue via an ester bond or an alkyl residue via an ether bond, whereas the sn-2 position almost exclusively contains acyl residues. The highly oxidized (n-3 and n-6 polyunsaturated fatty acids) are preferably bound to the sn-2 position of glycerophospholipids. Thus, most of the oxidized phospholipids are modified at this position. At the sn-1 position of glycerol a saturated fatty acid is frequently bound. Plasmalogens (alkenylacylglycerophospholipids) contain a vinyl ether bond in position sn-1 and, as a result, they are also susceptible to oxidative modifications at the sn-1 position.

The divergences in chemical structure of diverse types of phospholipids determine the physical properties of the membrane. PC tends to form bilayers with small curvature, while PE imposes a negative curvature on these lipid bilayers (Walton et al., 2003). Conversely, introduction of the micelle-forming LPC into a PC membrane results in a positive curvature. In addition to the polar head groups the polarity, length and unsaturation of the phospholipid acyl chains have also an impact on physical membrane properties. Thus, phospholipid oxidation products are very likely to change the properties of biological membranes, because their polarity and shape may differ considerably from the structures of their parent molecules. Thus, they may modify lipid–lipid and lipid–protein interactions and, as a consequence, also membrane protein functions.

When the sn-2 fatty acids of phospholipids are oxidized by radicals, numerous different types of oxidative products is formed (Catala, 2009). These comprise phospholipids containing fatty acid oxidation products (usually referred to as oxidized phospholipids), lysophospholipids, and fragmentation products of fatty acid oxidation. Some of these products, lysophosphatidic acid or lysophospholipids, can be formed both enzymatically and nonenzymatically. Work from several laboratories have identified a diversity of phospholipid oxidation products and confirmed bioactivities of these phospholipids on vascular wall cells, leukocytes, and platelets. Because of the great number of fatty acid oxidation products that have been identified, it is almost sure that many other bioactive oxidized phospholipids will be find out. There are several significant issues that confront investigators concerned in bioactive phospholipids, including preparation, identification, quantification, and biological checking.

Changes in phospholipid-induced oxidation reactions generates a large number of structurally dissimilar oxidation products, which difficult their isolation and characterization. Mass spectrometry (MS) and tandem mass spectrometry (MS/MS) using the soft ionization methods (electrospray and matrix-assisted laser desorption ionization) are the optimum approaches for the study of oxidized phospholipids. Product ions in tandem mass spectra of oxidized phospholipids allows identify changes in the fatty acyl chain and specific features such as existence of new functional groups in the molecule and their position along the fatty acyl chain (Lee et la., 1996).

Important Handmarks in the Knowledge of Polyunsaturated Fatty Acids in the Last Five Decades

In the last five decades, the significance of polyunsaturated fatty acids to biological systems and medicine began to be widely explored; the most important hand marks in the knowledge of polyunsaturated fatty acids in the last five decades are listed below:

1. Alexandri E, Ahmed R, Siddiqui H, Choudhary MI, Tsiafoulis CG, Gerothanassis IP. *High Resolution NMR Spectroscopy as a Structural and Analytical Tool for Unsaturated Lipids in Solution Molecules*. 2017; 22. pii: E1663. doi: 10.3390/molecules 22101663.

The objective of this review article is to provide: (i) an overview of the critical 1H-, 13C- and 31P-NMR parameters for structural and analytical investigations; (ii) an overview of various 1D and 2D NMR techniques that have been used for resonance assignments; (iii) selected analytical and structural studies with emphasis in the identification of major and minor unsaturated fatty acids in complex lipid extracts without the need for the isolation of the individual components; (iv) selected investigations of oxidation products of lipids; (v) applications in the emerging field of lipidomics; (vi) studies of protein-lipid interactions at a molecular level; (vii) practical considerations and (viii) an overview of future developments in the field.

2. Galano JM, Lee YY, Oger C, Vigor C, Vercauteren J, Durand T, Giera M, Lee JC. Isoprostanes, neuroprostanes and phytoprostanes: An overview of 25 years of research in chemistry and biology. *Prog Lipid Res.* 2017 Sep 18; 68: 83-108. doi: 10.1016/j.plipres.2017.09.004.

Since the beginning of the 1990's diverse types of metabolites originating from polyunsaturated fatty acids, formed under autooxidative conditions were discovered. Known as prostaglandin isomers (or isoprostanoids) originating from arachidonic acid, neuroprostanes from docosahexaenoic acid, and phytoprostanes from α-linolenic acid proved to be prevalent in biology. The syntheses of these compounds by organic chemists and the development of sophisticated mass spectrometry methods has boosted our understanding of the isoprostanoid biology. In recent years, it has become accepted that these molecules not only serve as markers of oxidative damage but also exhibit a wide range of bioactivities. In addition, isoprostanoids have emerged as indicators of oxidative stress in humans and their environment. This review explores in detail the isoprostanoid chemistry and biology that has been achieved in the past three decades.

3. Echeverría F, Valenzuela R, Catalina Hernandez-Rodas M, Valenzuela A. Docosahexaenoic acid (DHA), a fundamental fatty acid for the brain: New dietary sources Prostaglandins Leukot *Essent Fatty Acids*. 2017 124:1-10

Docosahexaenoic acid (C22: 6n-3, DHA) is a long-chain polyunsaturated fatty acid of marine origin fundamental for the formation and function of the nervous system, particularly the brain and the retina of humans. It has been proposed a remarkable role of DHA during human evolution, mainly on the growth and development of the brain. Currently, DHA is considered a critical nutrient during pregnancy and breastfeeding due their active participation in the development of the nervous system in early life. DHA and specifically one of its derivatives known as neuroprotectin D-1 (NPD-1), has neuroprotective properties against brain aging, neurodegenerative diseases and injury caused after brain ischemia-reperfusion episodes. This paper discusses the importance of DHA in the human brain given its relevance in the development of the tissue and as neuroprotective agent. It is also included a critical view about the ways to supply this noble fatty acid to the population.

4. Kuda O Bioactive metabolites of docosahexaenoic acid. *Biochimie.* 2017 136:12-20

Docosahexaenoic acid (DHA) is an essential fatty acid that is recognized as a beneficial dietary constituent and as a source of the anti-inflammatory specialized proresolving mediators (SPM): resolvins, protectins and maresins. Apart from SPMs, other metabolites of DHA also exert potent biological effects. This article summarizes current knowledge on the metabolic pathways involved in generation of DHA metabolites. Over 70 biologically active metabolites have been described, but are often discussed separately within specific research areas. This review follows DHA metabolism and attempts to integrate the diverse DHA metabolites emphasizing those with identified biological effects. DHA metabolites could be divided into DHA-derived SPMs, DHA epoxides, electrophilic oxo-derivatives (EFOX) of DHA, neuroprostanes, ethanolamines, acylglycerols, docosahexaenoyl amides of amino acids or neurotransmitters, and branched DHA esters of hydroxy fatty acids. These bioactive metabolites have pleiotropic effects that include augmenting energy expenditure, stimulating lipid catabolism, modulating the immune response, helping to resolve inflammation, and promoting wound healing and tissue regeneration. As a result they have been shown to exert many beneficial actions: neuroprotection, anti-hypertension, anti-hyperalgesia, anti-arrhythmia, anti-tumorigenesis etc. Given the chemical structure of DHA, the number and geometry of double bonds, and the panel of enzymes metabolizing DHA, it is also likely that novel bioactive derivatives will be identified in the future.

5. Yakubenko VP, Byzova TV. Biological and pathophysiological roles of end-products of DHA oxidation *Biochim Biophys Acta.* 2017, 1862: 407-415

Polyunsaturated fatty acids (PUFA) are known to be present and/or enriched in vegetable and fish oils. Among fatty acids, n-3 PUFA are generally considered to be protective in inflammation-related diseases. The guidelines for substituting saturated fatty acids for PUFAs have been highly publicized for decades by numerous health organizations. Recently, however, the beneficial properties of n-3 PUFA are questioned by detailed analyses of multiple randomized controlled clinical trials. The reported heterogeneity of results is likely due not only to differential effects of PUFAs on various pathological processes in humans, but also to the wide spectrum of PUFA's derived products generated in vivo.

6. Hishikawa D, Valentine WJ, Iizuka-Hishikawa Y, Shindou H, Shimizu T. Metabolism and functions of docosahexaenoic acid-containing membrane glycerophospholipids. *FEBS Lett*. 2017 591: 2730-2744

Omega-3 (ω-3) fatty acids (FAs) such as docosahexaenoic acid (DHA) and eicosapentaenoic acid (EPA) are known to have important roles in human health and disease. Besides being utilized as fuel, ω-3 FAs have specific functions based on their structural characteristics. These functions include serving as ligands for several receptors, precursors of lipid mediators, and components of membrane glycerophospholipids (GPLs). Since ω-3 FAs (especially DHA) are highly flexible, the levels of DHA in GPLs may affect membrane biophysical properties such as fluidity, flexibility, and thickness. Here, we summarize some of the cellular mechanisms for incorporating DHA into membrane GPLs and propose biological effects and functions of DHA-containing membranes of several cell and tissue types.

7. Yakubenko VP, Byzova TV. Biological and pathophysiological roles of end-products of DHA oxidation. *Biochim Biophys Acta.* 2017 1862: 407-415

The goal of this review is to discuss the studies focused on well-defined end-products of PUFAs oxidation, their generation, presence in various pathological and physiological conditions, their biological activities and known receptors. Carboxyethylpyrrole (CEP), a DHA-derived oxidized product, is especially emphasized due to recent data demonstrating its pathophysiological significance in many inflammation-associated diseases, including atherosclerosis, hyperlipidemia, thrombosis, macular degeneration, and tumor progression. CEP is a product of radical-based oxidation of PUFA that forms adducts with proteins and lipids in blood and tissues, generating new powerful ligands for TLRs and scavenger receptors. The interaction of CEP with these receptors affects inflammatory response, angiogenesis, and wound healing

8. Martin SA, Brash AR, Murphy RC. The discovery and early structural studies of arachidonic acid. *J Lipid Res.* 2016 57: 1126-1132

Arachidonic acid and esterified arachidonate are ubiquitous components of every mammalian cell. This polyunsaturated fatty acid serves very important biochemical roles, including being the direct precursor of bioactive lipid mediators such as prostaglandin and leukotrienes. This 20 carbon fatty acid with four double bonds was first isolated and identified from mammalian tissues in 1909 by Percival Hartley. This was accomplished prior to the advent of chromatography or any spectroscopic methodology (MS, infrared, UV, or NMR). The name, arachidonic, was suggested in 1913 based on its relationship to the well-known arachidic acid (C20:0). It took until 1940 before the positions of the four double bonds were defined at 5,8,11,14 of the 20-carbon chain. Total synthesis was reported in 1961 and, finally, the configuration of the double bonds was confirmed as all-cis-5,8,11,14. By the 1930s, the relationship of arachidonic acid within the family of essential fatty acids helped cue an understanding of its structure and the biosynthetic pathway. Herein, we review the findings leading up to the discovery of arachidonic acid and the progress toward its complete structural elucidation.

9. Calder PC. Marine omega-3 fatty acids and inflammatory processes: Effects, mechanisms and clinical relevance. *Biochim Biophys Acta.* 2015; 1851:469-484

Inflammation is a condition which contributes to a range of human diseases. It involves a multitude of cell types, chemical mediators, and interactions. Eicosapentaenoic acid (EPA) and docosahexaenoic acid (DHA) are omega-3 (n-3) fatty acids found in oily fish and fish oil supplements. These fatty acids are able to partly inhibit a number of aspects of inflammation including leukocyte chemotaxis, adhesion molecule expression and leukocyte-endothelial adhesive interactions, production of eicosanoids like prostaglandins and leukotrienes from the n-6 fatty acid arachidonic acid, production of inflammatory cytokines, and T-helper 1 lymphocyte reactivity. In addition, EPA gives rise to eicosanoids that often have lower biological potency than those produced from arachidonic acid and EPA and DHA give rise to anti-inflammatory and inflammation resolving mediators called resolvins, protectins and maresins. Mechanisms underlying the anti-inflammatory actions of marine n-3 fatty acids include altered cell membrane phospholipid fatty acid composition, disruption of lipid rafts, inhibition of activation of the pro-inflammatory transcription factor nuclear factor kappa B so reducing expression of inflammatory genes, activation of the anti-inflammatory transcription factor peroxisome proliferator activated receptor γ and binding to the G protein coupled receptor GPR120. These mechanisms are interlinked, although the full extent of this is not yet elucidated. Animal experiments demonstrate benefit from marine n-3 fatty acids in models of rheumatoid arthritis (RA), inflammatory bowel disease (IBD) and asthma. Clinical trials of fish oil in RA demonstrate benefit, but clinical trials of fish oil in IBD and asthma are inconsistent with no overall clear evidence of efficacy. This article is part of a Special Issue entitled "Oxygenated metabolism of PUFA: analysis and biological relevance."

10. Gong Y, Wan X, Jiang M, Hu C, Hu H, Huang F. Metabolic engineering of microorganisms to produce omega-3 very long-chain polyunsaturated fatty acids. *Prog Lipid Res*. 2014; 56:19-35

Omega-3 long-chain polyunsaturated fatty acids (LC-PUFAs) have received growing attention due to their significant roles in human health.

Currently the main source of these nutritionally and medically important fatty acids is marine fish, which has not met ever-increasing global demand. Microorganisms are an important alternative source also being explored. Although many microorganisms accumulate omega-3 LC-PUFAs naturally, metabolic engineering might still be necessary for significantly improving their yields. Here, we review recent research involving the engineering of microorganisms for production of omega-3 LC-PUFAs, including eicospentaenoic acid and docosohexaenoic acid. Both reconstitution of omega-3 LC-PUFA biosynthetic pathways and modification of existing pathways in microorganisms have demonstrated the potential to produce high levels of omega-3 LC-PUFAs. However, the yields of omega-3 LC-PUFAs in host systems have been substantially limited by potential metabolic bottlenecks, which might be caused partly by inefficient flux of fatty acid intermediates between the acyl-CoA and different lipid class pools. Although fatty acid flux in both native and heterologous microbial hosts might be controlled by several acyltransferases, evidence has suggested that genetic manipulation of one acyltransferase alone could significantly increase the accumulation of LC-PUFAs. The number of oleaginous microorganisms that can be genetically transformed is increasing, which will advance engineering efforts to maximize LC-PUFA yields in microbial strains.

11. Catalá A. Five decades with polyunsaturated Fatty acids: chemical synthesis, enzymatic formation, lipid peroxidation and its biological effects. *J Lipids*. 2013; doi: 10.1155/2013/710290

I have been involved in research on polyunsaturated fatty acids since 1964 and this review is intended to cover some of the most important aspects of this work. Polyunsaturated fatty acids have followed me during my whole scientific career and I have published a number of studies concerned with different aspects of them such as chemical synthesis, enzymatic formation, metabolism, transport, physical, chemical, and catalytic properties of a reconstructed desaturase system in liposomes, lipid peroxidation, and their effects. The first project I became involved in was the organic synthesis of (1-(14)C) eicosa-11,14-dienoic acid, with the aim of demonstrating the participation of that compound as a possible intermediary in the biosynthesis of arachidonic acid *"in vivo."* From 1966 to 1982, I was involved in several projects that study the metabolism of polyunsaturated fatty acids. In the eighties, we studied fatty acid binding protein. From 1990 up to now, our laboratory has been interested in the lipid peroxidation of biological membranes from various tissues and different species as well as liposomes prepared with phospholipids rich in PUFAs. We tested the effect of many antioxidants such as alpha tocopherol, vitamin A, melatonin and its structural analogues, and conjugated linoleic acid, among others.

12. Afman LA, Müller M. Human nutrigenomics of gene regulation by dietary fatty acids. *Prog Lipid Res*. 2012; 51:63-70

Nutrigenomics employs high-throughput genomics technologies to unravel how nutrients modulate gene and protein expression and ultimately influence cellular and organism metabolism. The most often-applied genomics technique so far is transcriptomics, which allows quantifying genome-wide changes in gene expression of thousands of genes at the same time in one sample. The performance of gene expression quantification requires sufficient high-quality homogenous cellular material, therefore research in healthy volunteers is restricted to biopsies from easy accessible tissues such as subcutaneous adipose tissue, skeletal muscle and intestinal biopsies or even more easily accessible cells such as peripheral blood mononuclear cells from blood. There is now significant evidence that fatty acids, in particular unsaturated fatty acids, exert many of their effects through modulation of gene transcription by regulating the activity of numerous transcription factors, including nuclear receptors such as peroxisome proliferator activated receptors, liver X receptor and sterol regulatory binding proteins. This review evaluates the human nutrigenomics studies performed on dietary fat since the initiation of nutrigenomics research around 10 years ago. Although the number of studies is still limited, all studies clearly suggest that changes in dietary fatty acids intake and composition can have a significant impact on cellular adaptive response capacity by gene transcription changes in humans. This adds important knowledge to our understanding of the strong effects that various fatty acids can have on numerous metabolic and inflammatory pathways, signaling routes and homeostatic control in the cell and ultimately on whole body health. It is important to use and integrate nutrigenomics in all future nutrition studies to build up the necessary framework for evidence-based nutrition in near future.

13. Kaur G, Cameron-Smith D, Garg M, Sinclair AJ. Docosapentaenoic acid (22:5n-3): a review of its biological effects. *Prog Lipid Res*. 2011 50: 28-34

This article summarizes the current knowledge available on metabolism and the biological effects of n-3 docosapentaenoic acid (DPA). n-3 DPA has not been extensively studied because of the limited availability of the pure compound. n-3 DPA is an elongated metabolite of EPA and is an intermediary product between EPA and DHA. The literature on n-3 DPA is limited, however the available data suggests it has beneficial health effects. In vitro n-3 DPA is retro-converted back to EPA, however it does not appear to be readily metabolised to DHA. In vivo studies have shown limited conversion of n-3 DPA to DHA, mainly in liver, but in addition retro-conversion to EPA is evident in a number of tissues. n-3 DPA can be metabolised by lipoxygenase, in platelets, to form ll-hydroxy-7,9,13,16,19- and 14-hydroxy-7,10,12,16,19-DPA. It has also been reported that n-3 DPA is effective (more so than EPA and DHA) in inhibition of aggregation in platelets obtained from rabbit blood. In addition, there is evidence that n-3 DPA possesses 10-fold greater endothelial cell migration ability than EPA, which is important in wound-healing processes. An in vivo study has reported that n-3 DPA reduces the fatty acid synthase and malic enzyme activity levels in n-3 DPA-supplemented mice and these effects were stronger than the EPA-supplemented mice. Another recent in vivo study has reported that n-3 DPA may have a role in attenuating age-related decrease in spatial learning and long-term potentiation. However, more research remains to be done to further investigate the biological effects of this n-3 VLCPUFA.

14. Kim W, Khan NA, McMurray DN, Prior IA, Wang N, Chapkin RS. Regulatory activity of polyunsaturated fatty acids in T-cell signaling. *Prog Lipid Res*. 2010; 49: 250-261

n-3 Polyunsaturated fatty acids (PUFA) are considered to be authentic immunosuppressors and appear to exert beneficial effects with respect to certain immune-mediated diseases. In addition to promoting T-helper 1 (Th1) cell to T-helper 2 (Th2) cell effector T-cell differentiation, n-3 PUFA may also exert anti-inflammatory actions by inducing apoptosis in Th1 cells. With respect to mechanisms of action, effects range from the modulation of membrane receptors to gene transcription via perturbation of a number of second messenger cascades. In this review, the putative targets of anti-inflammatory n-3 PUFA, activated during early and late events of T-cell activation will be discussed. Studies have demonstrated that these fatty acids alter plasma membrane micro-organization (lipid rafts) at the immunological synapse, the site where T-cells and antigen-presenting cells (APC) form a physical contact for antigen initiated T-cell signaling. In addition, the production of diacylglycerol and the activation of different isoforms of protein kinase C (PKC), mitogen-activated protein kinase (MAPK), calcium signaling, and nuclear translocation/activation of transcriptional factors, can be modulated by n-3 PUFA. Advantages and limitations of diverse methodologies to study the membrane lipid raft hypothesis, as well as apparent contradictions regarding the effect of n-3 PUFA on lipid rafts will be critically presented.

15. Cunnane SC, Plourde M, Pifferi F, Bégin M, Féart C, Barberger-Gateau P. Fish, docosahexaenoic acid and Alzheimer's disease. *Prog Lipid Res.* 2009; 48: 239-256

Cognitive decline in the elderly, particularly Alzheimer's disease (AD), is a major socio-economic and healthcare concern. We review here the literature on one specific aspect of diet affecting AD, that of the omega3 fatty acids, particularly the brain's principle omega3 fatty acid - docosahexaenoic acid (DHA). DHA has deservedly received wide attention as a nutrient supporting both optimal brain development and for cardiovascular health. Our aim here is to critically assess the quality of the present literature as well as the potential of omega3 fatty acids to treat or delay the onset of AD. We start with a brief description of cognitive decline in the elderly, followed by an overview of well recognized biological functions of DHA. We then turn to epidemiological studies, which are largely supportive of protective effects of fish and DHA against risk of AD. However, biological studies, including blood and brain DHA analyses need careful interpretation and further investigation, without which the success of clinical trials with DHA may continue to struggle. We draw attention to some of the methodological issues that need resolution as well as an emerging mechanism that may explain how DHA could be linked to protecting brain function in the elderly

16. Schmitz G, Ecker J. The opposing effects of n-3 and n-6 fatty acids. *Prog Lipid Res*. 2008; 47: 147-155

Polyunsaturated fatty acids (PUFAs) can be classified in n-3 fatty acids and n-6 fatty acids, and in westernized diet the predominant dietary PUFAs are n-6 fatty acids. Both types of fatty acids are precursors of signaling molecules with opposing effects, that modulate membrane microdomain composition, receptor signaling and gene expression. The predominant n-6 fatty acid is arachidonic acid, which is converted to prostaglandins, leukotrienes and other lipoxygenase or cyclooxygenase products. These products are important regulators of cellular functions with inflammatory, atherogenic and prothrombotic effects. Typical n-3 fatty acids are docosahexaenoic acid and eicosapentaenoic acid, which are competitive substrates for the enzymes and products of arachidonic acid metabolism. Docosahexaenoic acid- and eicosapentaenoic acid-derived eicosanoids antagonize the pro-inflammatory effects of n-6 fatty acids. n-3 and n-6 fatty acids are ligands/modulators for the nuclear receptors NFkappaB,

PPAR and SREBP-1c, which control various genes of inflammatory signaling and lipid metabolism. n-3 Fatty acids down-regulate inflammatory genes and lipid synthesis, and stimulate fatty acid degradation. In addition, the n-3/n-6 PUFA content of cell and organelle membranes, as well as membrane microdomains strongly influences membrane function and numerous cellular processes such as cell death and survival.

17. Jakobsson A, Westerberg R, Jacobsson A. Fatty acid elongases in mammals: their regulationn and roles in metabolism. *Prog Lipid Res.* 2006; 45: 237-249

A significant amount of the fatty acids synthesized by the cytosolic enzyme complex fatty acid synthase (FAS) or taken up by the diet are further elongated into very long chain fatty acids (VLCFA) in a four-step reaction cycle by membrane-bound enzymes predominantly located in the endoplasmic reticulum. Members of the Elovl (elongation-of-very-long-chain-fatty acids) gene family encode for enzymes (elongases), which are believed to perform the first, regulatory, step (condensation) in the elongation cycle in mammals. The family of enzymes consists of at least six members in mouse and human, believed to carry out substrate-specific elongation with fatty acids of different lengths and degrees of unsaturation. The ability to synthesize VLCFA is a ubiquitous system found in different organs and cell types. However, VLCFAs seldom occur unesterified. Instead, they are joined either by an ester or amide linkage to a broad variety of different lipid species. VLCFA are most commonly found as building blocks in sphingolipids, although they are also important constituents of glycerophospholipids, triacylglycerols, sterol- and wax-esters. To generalize, the fatty acid elongases can be divided into two major groups: (a) enzymes which are suggested to be involved in the elongation of saturated and monounsaturated VLCFA (ELOVL1, 3 and 6) and (b) enzymes which are elongases of polyunsaturated fatty acids (PUFA) (ELOVL2, 4 and 5). All the elongases exhibit specific spatial and temporal expression. In this review, we will present and discuss the regulation of the mammalian fatty acid elongases and their potential role in lipid metabolism. We will consider both the biochemical functions of the proteins, as well as their role in a more physiological context.

18. Leonard AE, Pereira SL, Sprecher H, Huang YS. Elongation of long-chain fatty acids. *Prog Lipid Res.* 2004; 43: 36-54

The identification of different elonagase and desaturase genes involved in the EPA/DHA production has important biotechnological applications. These genes can be used in the production of PUFA-rich transgenic plant oils for therapeutic and prophylactic use. Also, advances in understanding gene regulation in PUFA biosynthesis will also impact the single-cell oil industry, such that growth conditions of the microalgae can be manipulated to enhance the production of EPA/DHA. This in turn will impact the marine fish-farming industry which depends on microalgae for enhancing the levels of PUFAs in fish. All the above will eventually afford the public an economical source of balanced n-3/n-6 PUFA-enriched oils that will greatly impact general health and nutrition in the near future.

19. Jump DB. *J Biol Chem.* 2002 277: 8755-8758. The biochemistry of n-3 polyunsaturated fatty acids

Dietary n-3 polyunsaturated fatty acids (n-3 PUFA) have effects on diverse physiological processes impacting normal health and chronic disease, such as the regulation of plasma lipid levels, cardiovascular and immune function, insulin action, and neuronal development and visual function. Ingestion ofn-3 PUFA will lead to their distribution to virtually every cell in the body with effects on membrane composition and function, eicosanoid synthesis, and signaling as well as the regulation of gene expression. However, cell-specific lipid metabolism as well as the expression of fatty acid-regulated transcription factors likely plays an important role in determining how cells respond to changes in PUFA composition. In this mini review Jump highlight some of the recent advances in our understanding of n-3 PUFA effects on cells with an emphasis on those mechanisms likely to have a broad physiological impact.

20. Grimaldi PA Fatty acid regulation of gene expression. *Curr Opin Clin Nutr Metab Care.* 2001 5: 433-437

Over the past 10 years it has become evident that fatty acids regulate cellular functions by modulating gene expression. Fatty acids and fatty acid metabolites exert some of their effects on gene expression by affecting the activity of nuclear transcription factors, peroxisome proliferator-activated receptors and sterol regulatory element binding protein type 1. The present review describes the latest developments in the field, with particular emphasis on the physiological roles of the various peroxisome proliferator-activated receptor isotypes, including their implication in the control of proliferation and differentiation of normal and malignant cells, and on the mechanisms implicated in the regulation of sterol regulatory element binding protein type 1 activity by polyunsaturated fatty acids.

21. Spector AA, Yorek MA. Membrane lipid composition and cellular function. *J Lipid Res.* 1985; 26: 1015-1035

The effects of lipid modification on cellular function are very complex. They often vary from one type of cell to another, and they do not exert a uniform effect on all processes in a single cell line. Therefore, it is not yet possible to make any generalizations or to predict how a given system will respond to a particular type of lipid modification. Many of the functional responses probably are caused directly by the membrane lipid structural changes, which affect either bulk lipid fluidity or specific lipid domains. The conformation or quaternary structures of certain transporters, receptors, and enzymes probably are sensitive to changes in the structure of their lipid microenvironment, leading to changes in activity.

22. Stubbs CD, Smith AD. The modification of mammalian membrane polyunsaturated fatty acid composition in relation to membrane fluidity and function. *Biochim Biophys Acta.* 1984; 779: 89-137

In this introductory section our objective has been to provide a background against which the more detailed discussions which follow may be set. We have abstracted the important points that relate specifically to unsaturation, and have given references to other reviews which cover the topics in a more general sense. In particular, we have attempted here to establish the meaning of a number of terms which are subsequently widely used.

23. Brenner RR. Effect of unsaturated acids on membrane structure and enzyme kinetics. *Prog Lipid Res.* 1984; 23:69-96

This review tries to show that, in spite of a rapid development and application of different physiochemical tools and their addition to the biochemical paraphernalia, the investigation of lipid-protein interactions is only beginning to approach the problem at the molecular level. Most of the information on the molecular structure of lipid bilayers and their changes with the corresponding modifications of the intrinsic protein behavior is lacking or is not precise. We even still do not know in which place of the membrane bilayer many of the proteins investigated are located. The lipid-protein interaction on the membranes is a biophysico chemical and, in many cases, vectorial problem that can only be solved by the application of physico-chemical and biochemical techniques joined by a common objective.

24. Wolfe LS. Eicosanoids: prostaglandins, thromboxanes, leukotrienes, and other derivatives of carbon-20 unsaturated fatty acids. *J Neurochem.* 1982; 38:1-14

In recent years, knowledge of the biochemistry of oxygenated metabolites of arachidonic acid has greatly increased. Their biological functions in acceleration and prevention of platelet aggregation and in inflammatory and immune reactions are becoming much clearer. The therapeutic value, particularly of PGI2 as well as selective inhibitors of synthesis, is also rapidly advancing. Despite much effort, the functional importance of prostaglandins and thromboxanes in the cNS in normal ongoing physiological processes is still quite uncertain. However, when parenchymal or vascular elements are damaged or invaded by extraneural cells, the synthesis of one or the other member of the eicosanoids is greatly increased and contributes significantly to pathophysiological reactions. Thus, prevention of synthesis is likely to have increasing importance in clinical neurology, particularly in cerebrovascular diseases.

25. Hamilton JG, Karol RJ. High performance liquid chromatography (HPLC) of arachidonic acid metabolites *Prog Lipid Res*. 1982; 21:155-170

Many of the steps in the metabolic pathway for arachidonic acid (Figure 1) were elucidated using the conventional chromatographic techniques of open column, thin layer and gas-liquid chromatography (GLC). However, these techniques are now being replaced by high performance liquid chromatography (HPLC). In fact, the characterization of prostacyclin and the leukotrienes were dependent on fast, efficient HPLC separations. The following is an overview of arachidonic acid metabolism with emphasis on the use of HPLC for the isolation, characterization and quantitation of prostaglandins, thromboxanes, hydroxy fatty acids and the leukotrienes.

26. Robison WG1, Kuwabara T, Bieri JG. The roles of vitamin E and unsaturated fatty acids in the visual process. *Retina.* 1982; 2: 263-281

Relatively high proportions of long-chain, polyunsaturated fatty acids seem to be required in rod photoreceptor membranes in order to provide the precise microenvironment for the proper function of the visual pigment rhodopsin. At the same time, such high levels of lipid unsaturation put the photoreceptor membranes at a high risk for autoxidation. The antioxidant vitamin E which can minimize autoxidation of polyunsaturated fatty acids is found in rather high concentrations in the outer segment membranes. Dietary deficiency in vitamin E induces disintegration of rod outer segment membranes, probably by increasing autoxidation. Also, it greatly accelerates the accumulation of aging pigments in the retinal pigment epithelium, probably because these lipofuscin granules do indeed represent the end products of lipid peroxidation. Vitamin E supplements, up to threefold normal levels, appear to provide no significant protection of the retina from light damage produced either by short but acute or by long-term, low level exposures to light. This is not consistent with current theories which implicate lipid peroxidation in the destruction of rod outer segments in light damaged retinas; more work is needed before any relation between retinal light damage and vitamin E levels can be assessed.

27. Cornwell DG, Panganamala RV. Atherosclerosis: an intracellular deficiency in essential fatty acids. *Prog Lipid Res*. 1981; 20: 365-376

Essential fatty acids are shunted into an intracellular cholesteryl ester pool when the endothelium is injured. An intracellular essential fatty acid deficiency occurs. The consequences of this process are enormous. Less arachidonic acid is available for the synthesis of prostacyclin enhancing both endothelial injury and platelet aggregation. Polyunsaturated fatty acids are not available for the control of cell proliferation through the oxidant stress which is induced by hydroperoxy fatty acids. Smooth muscle cells proliferate and the atherosclerotic lesion develops.

28. Yamanaka WK, Clemans GW, Hutchinson ML. Essential fatty acids deficiency in humans. *Prog Lipid Res.* 1980; 19: 187-215.

A need remains for further research concerning the desired level of EFA intake for normal cell functioning in healthy individuals and in those suffering from disease. Continued development of indices used to identify optimal plasma and cellular EFA levels also seems in order. Recent research indicates that EFA functions and requirements are greatly influenced by various minerals and by protein intake; these findings need to be expanded and further clarified.

29. Schauenstein E. Autoxidation of polyunsaturated esters in water: chemical structure and biological activity of the products. *J Lipid Res.* 1967 5: 417-428

When polyunsaturated esters or fatty acids are dispersed for long periods in water in the presence of air, water-soluble substances are formed in great variety. Because these short-chain products are constantly eluted by the aqueous phase and are consequently not available for further reaction in the oil phase, many intermediates of classical autoxidation can be isolated and identified. The identification of several of these compounds is described. Some of the peroxidic and nonperoxidic autoxidation products show biochemical activity-in particular, inhibition of glycolysis and of respiration during incubation with tumor cells in vitro. Minimal inhibitory concentrations of pure, isolated products have been determined for Ehrlich ascites tumor cells. Synthetic short-chain (C(4)-C(10)) hydroxylated alpha,-unsaturated aldehydes have been shown to have this action and also to cause morphological changes in these cells which quickly lead to their death. Normal cells are not affected. Possible therapeutic use of these compounds in the treatment of malignant neoplasms is discussed.

References

Bull H. 1899. Über die Bestimmung Stark Ungesättigter Fettsäuren in den Thranen. [About Determination of Highly Unsaturated Fatty Acids in Tears.] *Chemiker-Zeitung*. 23: 1043–1044.

Burr G. O. Significance of the essential fatty acids, *Fed. Proc.* 1 (1942) 224-233.

Burr G. O., Burr M. M. A new deficiency disease produced by the rigid exclusion of fat from the diet, *J. Biol. Chem.* 82 (1929) 345-367.

Burr G. O., Burr M. M. On the nature and role of the fatty acids essential in nutrition, *J. Biol. Chem.* 86 (1930) 587-621.

Catalá. A. Lipid peroxidation of membrane phospholipids generates hydroxy-alkenals and oxidized phospholipids active in physiological and/or pathological conditions. *Review Chem Phys Lipids* 157 (2009) 1-11.

Cole A. L., Subbanagounder G., Mukhopadhyay S., Berliner J. A., Vora D. K. Oxidized phospholipid-induced endothelial cell/monocyte interaction is mediated by a cAMP-dependent R-Ras/PI3-kinase pathway. *Arterioscler. Thromb. Vasc. Biol.* 23, 1384-1390, 2003.

Dale H. H. 1957. Percival Hartley 1881–1957. *Biogr. Mem. Fellows R. Soc.* 3: 80–100.

Green, T. G.; Hilditch, T. P. (1935). "The identification of linoleic and linolenic acids." *Biochem. J.* 29 : 1552–1563.

Greenberg ME, Sun M, Zhang R, Febbraio M, Silverstein R, Hazen SL. Oxidized phosphatidylserine-CD36 interactions play an essential role in macrophage-dependent phagocytosis of apoptotic cells. *J. Exp. Med.* 203, 2613-2625, 2006.

Gugiu B. G., Mesaros C. A., Sun M., Gu X., Crabb J. W., Salomon RG. Identification of oxidatively truncated ethanolamine phospholipids in retina and their generation from polyunsaturated phosphatidylethanolamines *Chem. Res. Toxicol.* 19, 262-271, 2006.

Knutzon D. S., Thurmond J. M., Huang Y. S., et al., Identification of Δ5-desaturase from Mortierella alpina by heterologous expression in bakers' yeast and canola. *J. Biol. Chem.* 1998; 273: 29360–29366.

Lee Y. C., Zheng Y. O., Taraschi T. F., Janes N. Hydrophobic alkyl headgroups strongly promote membrane curvature and violate the headgroup volume correlation due to "headgroup" insertion. *Biochemistry* 35, 3677-3684, 1996.

Marszalek J. R., Lodish H. F. Docosahexaenoic acid, fatty acid-interacting proteins, and neuronal function: breastmilk and fish are good for you. *Annu Rev Cell Dev Biol.* 2005; 21: 633–657.

Maskrey B. H., Bermúdez-Fajardo A., Morgan A. H., Stewart-Jones E., Dioszeghy V., Taylor G. W., Baker P. R., Coles B., Coffey M. J., Kühn H, O'Donnell V. B. Activated platelets and monocytes generate four hydroxyphosphatidylethanolamines via lipoxygenase. *J. Biol. Chem.* 282, 20151-20163, 2007.

Matsura T., Togawa A., Kai M., Nishida T., Nakada J., Ishibe Y., Kojo S., Yamamoto Y., Yamada K. The presence of oxidized phosphatidylserine on Fas-mediated apoptotic cell surface *Biochim. Biophys. Acta,* 1736, 181-188, 2005.

McCutcheon J. W. (1955). "Linolenic acid." *Org. Synth.; Coll.* Vol., 3, p. 351.

Rollett, A. (1909). "Zur kenntnis der linolensäure und des leinöls." ["To know linolenic acid and linseed oil."] *Z. Physiol. Chem.* 62 (5–6): 422–431.

Stedmans Medical Dictionary 26th Edition – 1926.

Tsujimoto M. 1906. On a new unsaturated fatty acid in japanese sardine oil. *J. College Engineering Tokyo.* 4: 1–10.

Walton K. A., Cole A. L., Yeh M., Subbanagounder G., Krutzik S. R., Modlin R. L., Lucas R. M., Nakai J., Smart E. J., Vora D. K., Berliner J. A. Specific phospholipid oxidation products inhibit ligand activation of toll-like receptors 4 and 2. *Arterioscler. Thromb. Vasc. Biol.* 23, 1197-1203, 2003.

In: Polyunsaturated Fatty Acids (PUFAs)
Editor: Angel Catala
ISBN: 978-1-53613-572-5

Chapter 6

ARE POLYUNSATURATED FATTY ACIDS BENEFICIAL FOR FAILING HEARTS?

Stephen C. Kolwicz Jr.**, *PhD
Heart and Muscle Metabolism Laboratory,
Health and Exercise Physiology Department,
Ursinus College, Collegeville, PA, US

ABSTRACT

Dietary consumption of polyunsaturated fatty acids (PUFAs), particularly when replacing saturated fatty acids, is a current behavioral modification strategy recommended by the American Heart Association/American College of Cardiology for the treatment and prevention of cardiovascular diseases. There is strong epidemiological data showing that both cardiovascular risk and mortality is reduced when PUFAs are consumed. Although PUFAs have demonstrable benefits on serum lipid profiles, less is known about the direct effects on myocardial function, particularly in heart failure. In this review, a summary of the recent literature on the outcomes of PUFA consumption or supplementation on measures of cardiac function in both clinical and

* Corresponding Author: skolwicz@ursinus.edu.

animal models of heart failure will be presented. As PUFAs are essential components of membranes, dietary modulation of membrane lipid composition, especially of the mitochondrial membrane, and its functional consequences will be explored. Finally, the role of PUFAs as signaling molecules to modulate both inflammatory and fibrosis pathways will be discussed.

INTRODUCTION

Despite generally considered as a nutrient detrimental to overall health, fatty acids are critical substrates necessary for a variety of physiological functions in the heart. First, fatty acids are an essential nutrient for energy production, accounting for more than half of ATP regeneration in the healthy heart while storage of fatty acids in triacylglycerol represents a reservoir of substrates for ATP synthesis [1, 2]. Fatty acids are also important biosynthetic substrates allowing for the formation of phospholipid membranes of the cell and organelles such as the endoplasmic reticulum and mitochondria [3]. Finally, fatty acids also function as signaling molecules that affect a myriad of pathways including apoptosis, endoplasmic reticulum stress, and autophagy [4-6]. Therefore, intake of fatty acids are required for overall health and homeostasis of the myocardium.

Biochemically, fatty acids are carboxylic acids with long aliphatic chains that may or may not contain carbon-to-carbon double bonds. The absence of double bonds defines the particular fatty acid as "saturated" whereas the presence of double bonds defines the fatty acid as "unsaturated." The biochemical structure of these individual fatty acids can have profound biological and physiological consequences that will be discussed in more detail within this chapter. These consequences are clearly demonstrated in the epidemiological literature where high intakes of saturated fatty acids are often associated with increased risk and mortality from cardiovascular diseases while high intakes of unsaturated fatty acids are associated with reduced risk and mortality [7]. Therefore, not all fatty acids present in food sources are equal in their ability to elicit physiological or pathological cellular actions.

Unsaturated fatty acids are generally considered the "healthy" fatty acids, containing one or more carbon-to-carbon double bonds. However, the variety of the number of double bonds present in the structure offers additional classification. One double bond or "site of unsaturation" classifies the fatty acid as "monounsaturated" while more than one site of unsaturation classifies the fatty acid as "polyunsaturated." A plethora of research has debated whether higher dietary consumption of monounsaturated or polyunsaturated fatty acids are better for overall health and mortality. Certainly, the Mediterranean diet, which results in a higher consumption of monounsaturated fatty acids found in olive oil, has been recommended for reduced risk of heart disease [8], diabetes [9], and cancer [10]. Additionally, consumption of fish, oils, and nuts, foods high in polyunsaturated fatty acids, are also beneficial to overall health and lowers risk of disease and, are therefore, recommended by several professional health entities [7, 11, 12]. The American Heart Association currently recommends replacing dietary consumption of saturated fatty acids with both mono- and poly-unsaturated fatty acids [13]. This chapter will focus specifically on the effects that polyunsaturated fatty acids has on cardiac function.

WHAT ARE POLYUNSATURATED FATTY ACIDS (PUFAS)?

Polyunsaturated fatty acids (PUFAs) contain more than one carbon-to-carbon double bounds in their biochemical structure. These double bonds are typically in the *cis* isomer form, meaning that the hydrogen atoms are located on the same side as the double bond. (In the *trans* isomer, the hydrogen atoms are found on the opposite side of the double bond are a generally found in low abundance naturally). PUFAs are further classified as Omega-3 or Omega-6 fatty acids based on the precise location of the double bonds. The carboxyl group signifies the beginning of the fatty acid chain and the carbon immediately after that is denoted as the "alpha" carbon. The last carbon in the chain is referred to as the "omega" carbon. Therefore, if a double bond is present between the third and fourth carbon from the omega end, it is called "Omega-3" while the double bond being present between the sixth and seventh carbon from the omega end is called "Omega-6."

Nomenclature of PUFAs

Both Omega-3 and Omega-6 fatty acids are subject to shorthand nomenclature that provides additional description based on the biochemical structure. First, these fatty acids can be described as intermediate-chain (i.e., 18 carbons or less) or long-chain (i.e., greater than 20 carbons). Second, the exact number of double bonds present in the structure is also referenced. For further description, the series n-3 (omega-3) or n-6 (omega-6) is also provided. Therefore, linoleic acid, an intermediate-chain omega-6 PUFA of 18 carbons and 2 double bonds has a lipid number of C18:2n-6. Table 1 presents the major PUFAs with the associated lipid number.

Fatty acids are also described as being essential or non-essential depending on the capacity for cellular synthesis. Thus, essential fatty acids cannot be created via biochemical synthesis and must be obtained from outside sources. The intermediate-chain omega-3 and omega-6 PUFAs, alpha-linolenic acid (ALA) and linoleic acid (LA), respectively, are the only two fatty acids deemed essential for human physiology. Therefore, exogenous intake from food sources is paramount for survival. Once ingested, both ALA and LA can be converted to other relevant PUFAs through a complex series of reactions requiring several desaturase and elongase enzymes [14]. However, the capacity for these series of reactions is generally inefficient [15]. Therefore, exogenous consumption of long-chain PUFAs, particularly eicosapentaenoic acid (EPA, C20:5n-3) and docosahexaenoic acid, (DHA, C22:6n-3), is generally recommended to achieve relevant physiological concentrations.

Table 1. Polyunsaturated fatty acids and lipid number

Omega-3 Fatty Acid	Lipid Number	Omega-6 Fatty Acid	Lipid Number
Alpha-linolenic acid (ALA)	C18:3n-3	Linoleic Acid (LA)	C18:2n-6
Eicosapentaenoic acid (EPA)	C20:5n-3	Gamma-linolenic acid (GLA)	C18:3n-6
Docosahexaenoic acid (DHA)	C22:6n-3	Arachidonic acid (AA, ARA)	C20:4n-6

The major polyunsaturated fatty acids and their associated lipid number is presented. For the lipid number, C represents the number of carbons in the fatty acid change. After the colon, the number of carbon-to-carbon bonds is indicated. The site of the double bond from the fatty acid tail (omega end) indicate whether the fatty acid is an omega-3 (n-3) or omega-6 (n-6).

Reduction of Cardiovascular Disease with PUFAs

Much of the clinical and epidemiological literature has primarily focused on the specific benefits of omega-3 polyunsaturated fatty acids (n-3 PUFAs) on cardiovascular disease risk. In particular, the original population studies in the 1970s established a link between the consumption of fish and cardiovascular disease risk profile [16, 17]. The findings resulted in health organizations, including the American Heart Association, to encourage the consumption of n-3 PUFAs in the form of fish for both at risk individuals as well as those with existing cardiac disease [13]. Additional insight has been garnered from research in both human and animal models, demonstrating the effectiveness of n-3 PUFAs in reducing cellular inflammatory responses [18, 19], decreasing serum triglycerides [20], reducing lethal cardiac arrhythmias [21], and promoting vascular health [22]. However, controversy on the secondary prevention of n-3 PUFAs in patients with coronary heart disease or heart failure still exists and is discussed later in this chapter.

Dietary Intake and Supplementation with PUFAs

The epidemiological literature supports the concept that cardioprotective benefits are obtained in diets that consume higher n-3 PUFAs at the expense of saturated fatty acids. Consistent with this, major health organizations, including the American Heart Association (AHA), currently recommend the consumption of two portions of fatty fish per week resulting in a consumption of approximately 1 gram combined of the n-3 PUFAs, EPA and DHA [12, 13]. In fact, 2-4 grams of EPA and DHA in fish oil supplements with a physician prescription was previously recommended [23]. However, recent clinical trials do not generally find positive benefits on secondary prevention, particularly using supplementation strategies [24-28]. Several factors may contribute to the null findings in these studies including cofounding factors of other pharmacological agents, dosages, and treatment periods [29,30]. Despite the lack of clarity in the literature, approximately one-third of adults that regularly use dietary supplements, consume fish oil capsules that contain high quantities of EPA and DHA. However, this presents a potential issue since the quality of the individual supplement is often neglected at the expense of economic costs. A recent study found that only 3 of 32 fish oil supplements contained EPA and DHA levels that were consistent with their label contents [31]. Overall, the majority of the supplements had n-3 PUFA content that was 30% lower than the label claim. As PUFAs are particularly prone to oxidation, obtaining these through sources other than dietary foods presents a unique problem. In a recent study, approximately 80% of the supplements had higher than acceptable levels of lipid oxidation markers [31]. Although the physiological significance of these oxidative by-products are still yet unknown [32], their presence likely explains the reduced fatty acid contents in the supplement. In addition, the lower n-3 PUFA content in supplements might require greater consumption to reach recommended levels. Higher quality, higher dosage supplements are available on a prescription basis but are equated with higher costs [33]. Despite the lack of clarity in the literature, the AHA maintains its recommendations for n-3 PUFA supplementation in patients with coronary heart disease [34].

Do PUFAs Directly Affect the Myocardium?

The systemic effects of n-3 PUFAs and risk of cardiovascular events and mortality have been generally well described in the scientific literature. However, the direct benefits that n-3 PUFAs elicit on cardiac myocyte biology are less clear. There is the perception that n-6 PUFAs are more prone to activate pro-inflammatory signaling pathways while n-3 PUFAs augment anti-inflammatory signaling pathways [35]. This explains the preponderance of studies in both human and animal models that examine the effects of dietary or supplementation strategies of n-3 PUFAs in a variety of cardiovascular pathologies. However, more recently the serum ratios of n-6 to n-3 PUFAs have been advocated as a more relevant physiologic marker [36, 37].

Several clinical trials have reported the effects of n-3 PUFA treatment on cardiac function in human heart failure. The most well-known trial, the Gruppo Italiano per lo Studio della Sopravvivenza nell'Infarto miocardico heart failure cohort (GISSI-HF) found a small effect of n-3 PUFAs on mortality and cardiovascular admissions in patients with New York Heart Association (NYHA) class II-IV, without a significant effect on cardiac function [38]. However, some trials did find minor improvements in left ventricular ejection fraction (LVEF) [18, 25, 39], diastolic dysfunction [40], and/or exercise performance [18] in patients with varying heart failure etiologies. Unfortunately, other clinical trials (i.e., JELIS, DART2, Alpha-Omega, Omega, and SU.FOL.OM3) in patients with coronary heart disease have not supported the benefit of n-3 PUFAs on functional outcomes [24, 26-28, 41]. Experimental design, among other issues, have questioned the validity of the findings [29, 30]. Another key issue that will need to be addressed throughout future studies is to determine the effective therapeutic dosage range of protective benefits, a therapeutic index, which has been suggested by animal models [30].

Studies in animal models of cardiac dysfunction as well as in cell culture systems have suggested some potential molecular mechanisms by which n-3 PUFA might modulate pathological signaling, leading to beneficial cardiac outcomes. As shown in Figure 1, these research studies have focused on several pathological and physiological processes including: 1) myocardial inflammatory pathways; 2) ion channel physiology; 3) cardiac fibrosis; and 4) mitochondrial membrane composition. The major findings from the research literature in each of these categories will be discussed hereafter.

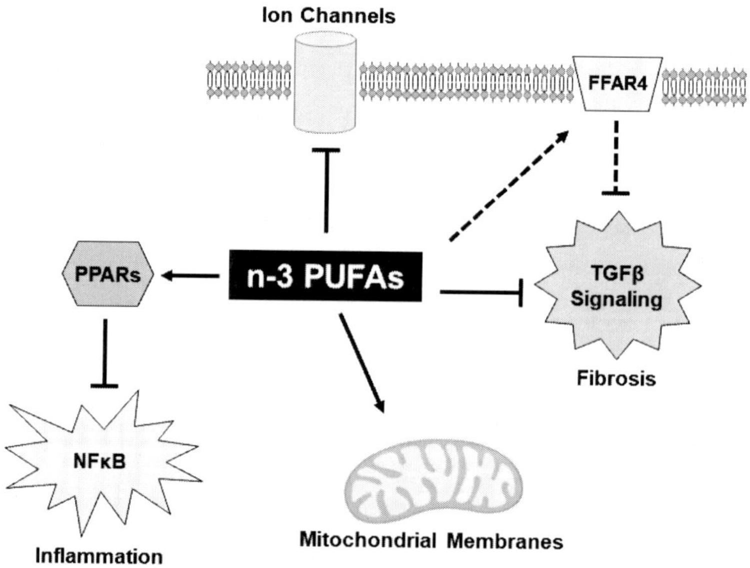

Figure 1. Potential mechanisms by which polyunsaturated fatty acids affect the cardiac myocyte.

Omega-3 polyunsaturated fatty acids (n-3 PUFAs) may benefit the myocardium during various disease pathologies by modulating several pathways. Evidence suggests that direct interaction with ion channels can stabilize membrane depolarization and decrease cardiac arrhythmias. Interaction of n-3 PUFAs with free fatty acid receptor 4 (FFAR4) may attenuate fibrosis via inhibition of transforming growth factor beta (TGFβ) signaling. Although n-3 PUFAs appear to influence the composition of the mitochondrial membrane, the effect on mitochondrial function is still unknown. Stimulation of peroxisome-proliferator activator receptor (PPAR) pathway can inhibit the inflammatory pathway through nuclear factor kappa (NFκB).

PUFAs and Myocardial Inflammation

Inflammation is a response necessary to maintain the physiological integrity of the cardiac structure during insult. Both acute injury and systemic inflammation can initiate the inflammatory mechanism within myocardial tissue. Heart failure has been considered a systemic inflammatory condition [42]. Activation of toll like receptor 4 (TLR4) is thought to be a critical trigger of inflammation contributing to heart failure [43]. Activation of the tissue necrosis factor (TNF) - nuclear factor kappa-light-chain-enhancer of activated B cells (NFκB) pathway also contributes to cardiomyopathy and heart failure [44, 45]. Therefore, the role of PUFAs in modulation of inflammatory pathways requires exploration.

PUFAs, especially the n-6 PUFA, arachidonic acid (AA), are thought to activate inflammatory pathways. The essential n-6 PUFA linoleic acid (LA) is converted to AA via the action of specific desaturase and elongase enzymes [14]. Subsequently, AA through the action of specific cyclooxygenases and lipoxygenases can be converted to prostaglandins, thromboxanes, and leukotrienes, collectively referred to as eicosanoids. These AA-derived eicosanoids are known regulators of a variety of inflammatory pathways, and thus, are considered pro-inflammatory. Recently, we observed an association between elevated serum arachidonic acid (AA) and diastolic dysfunction in mice fed a high fat diet that was predominantly unsaturated fatty acids [46]. In particular, the diet contained approximately one-third PUFAs, with an overwhelming amount of linoleic acid (C18:2n-6). Although not directly tested in our study [46], it is certainly reasonable to speculate that the high intake of LA resulted in elevated serum AA that may contribute to the activation of inflammatory pathways that resulted in diastolic dysfunction.

Similar to n-6 PUFAs, n-3 PUFAs, particularly EPA, also lead to the production of eicosanoids. The essential fatty acid, alpha-linolenic acid (ALA) is a precursor for EPA synthesis using the same desaturase and elongase enzymes that convert LA to AA. EPA can also produce eicosanoids but these generally modulate anti-inflammatory pathways. It is thought that the ratio of the n-6 to n-3 PUFA ratio is important as the competition between enzyme and substrate could ultimately lead to an altered balance of pro- and anti-inflammatory eicosanoids. Thus, a logical conclusion that n-6 PUFAs cause inflammation and n-3 PUFAs prevent inflammation is reached.

Research has shown that n-3 PUFAs are associated with reduced concentrations of TNF in cardiomyopathy patients [18]. However, recent clinical studies do not support the concept that intake of the LA does not equate with serum AA levels [47]. Moreover, increased LA does not lead to increase markers of inflammation [48]. It has been shown that EPA and DHA are potent activators of regulators of lipid metabolism, peroxisome proliferator-activated receptors (PPAR) [49]. As PPARs are also known to inhibit NFkB expression [50], the association of the anti-inflammatory action of n-3 PUFAs is theorized. The links between n-3 PUFAs and inflammation have been mostly observed in endothelial dysfunction and atherosclerosis [35]. Since little is known about the direct effects on the myocardium, future research could help determine a precise link in cardiac inflammatory pathways.

PUFAs and Cardiac Fibrosis

Following injury, a dynamic process involving deposition of connective tissue occurs in order preserve the myocardial structure. This process, termed fibrosis, is mediated by myofibroblast crosstalk to cardiomyocytes during cardiac remodeling. Based on the dynamic relationship between the inflammatory and fibrotic pathways in the heart [51], coupled with the potential for n-3 PUFAs to be cardioprotective [52], a new hypothesis regarding the ability of n-3 PUFAs to attenuate cardiac fibrosis has emerged as a potential molecular link.

Dietary supplementation of fish oil in mice for 12 weeks increased both red blood cell and left ventricular content of both EPA and DHA that led to a preservation of ejection fraction and reduced fibrosis in pressure-overload hypertrophy induced by transverse aortic constriction (TAC) [53]. Although transforming growth factor beta-1 (TGFβ1) upregulation and its downstream mediator, mothers against decapentaplegic homolog 2 (Smad2) and Smad3 phosphorylation were not prevented, both EPA and DHA blocked the nuclear translocation of the Smad proteins via an increase in the cyclic guanosine monophosphate/protein kinase G (cGMP/PKG) pathway [53]. This appeared to be mediated by an increase in nitric oxide (NO) production [53]. Because the doses of fish oil used in the previous study yielded supra-physiologic levels of n-3 PUFAs, the research group performed additional experiments to achieve n-3 PUFA levels consistent with clinical studies [54]. Using a reduced treatment strategy, they reported that although cardiac dysfunction was not prevented, EPA, but not DHA, was responsible for the reduction of fibrosis in mouse hearts subjected to TAC [55]. In their work, the authors suggested that the G-protein coupled free fatty acid receptor 4 (FFAR4) activation by EPA was responsible for the cardioprotective signaling [55]. FFARs, specifically FFAR1 and FFAR4, are activated by long chain PUFAs [56]. In this regard, n-3 PUFAs may potentially be cardioprotective in activating signaling cascades that inhibit pro-fibrotic pathways during disease states.

PUFAs, Ion Channels, and Cardiac Arrhythmias

One promising target by which PUFAs have a potential positive effect on the heart is by the modulation of ion channel action, stabilizing the excitability of the membrane and, thus, reducing aberrant cardiac rhythms. The GISSI-Prevention secondary prevention study concluded that fish oil capsules, combined with standard care, led to a significant reduction in sudden cardiac death [57]. In this regard, the ability of PUFAs to modulate ion channels and subsequent action potentials was proposed. Hypoxia enhances sodium currents that can active the reverse sodium-calcium exchanger, leading to elevations in both sodium and calcium that can induce cardiac arrhythmias. Basic science studies, including in-vitro cell and tissue preparations and animal studies, do demonstrate *the ability of* PUFAs (both omega-3 and omega-6) to modulate activity of sodium, potassium, and calcium channels as well as the sodium-calcium exchanger and calcium-activated potassium channels [58-61]. The mechanism of action is not quite clear from the literature as incorporation of arachidonic acid into cell membrane preparations was not observed [62] and, therefore, direct binding of these fatty acids to the specific channels is the likely mechanism.

In contrast, clinical trials within the last several years do not consistently support the findings observed in the previous clinical trials or basic science studies. Some interventional studies demonstrate that n-3 PUFA supplementation does not prevent ventricular arrhythmias or sudden cardiac death [26, 27] or reduce atrial fibrillation [63, 64]. In fact, in larger animal models, n-3 PUFA treatment had a proarrhythmic effect in inducing ventricular fibrillation during myocardial infarction in a canine model [65] or during ischemia in isolated pig hearts [66]. A similar increase in arrhythmogenicity was observed in patients with coronary heart disease [41]. Despite the wealth of studies, the precise mechanism by which PUFAs prevent or promote cardiac arrhythmias is still lacking. It is clear from the literature that a greater understanding of how PUFAs affect cardiac rhythm is needed before definitive conclusions can be made.

PUFAs and Mitochondrial Membranes

Phospholipids (PLs) are a major component of the mitochondria membrane, and thus, are crucial for overall integrity and function. The PL that has received the most attention in the literature is cardiolipin, which comprises ~20% of all PLs, and is found primarily in the inner mitochondrial membrane where it is associated with proteins of the electron transport chain [67, 68]. Cardiolipin is unique in that is predominantly composed of four molecules of the n-6 PUFA, linoleic acid (C18:2n-6) [67]. Research has suggested that a higher abundance of unsaturated FAs, especially PUFAs, within the mitochondrial PL membrane is important for maintaining membrane fluidity and function [69, 70]. An essential question is whether the composition of mitochondrial membrane changes during acute or chronic stress. With a relatively high abundance of PUFAs in the mitochondrial membrane, a loss of the PL integrity could lead to deficiencies in mitochondrial function.

There is considerable evidence demonstrating changes in the fatty acid composition of the mitochondrial membrane in both human and animal models of cardiac pathologies. Within these studies, a rather consistent pattern of phospholipid remodeling emerges, with a loss of linoleic acid (C18:2n-6) with a concomitant increase in either arachidonic acid (AA, C20:4n-6) or DHA (C22:6n-3) [71-74]. Several studies from the laboratory of the late William C. Stanley attempted to modulate the fatty acid composition of the mitochondrial membrane with n-3 PUFA nutritional strategies. Although increased dietary intake of EPA and/or DHA enhanced phospholipid content of these fatty acids while decreasing content of AA, the subsequent effects on cardiac and mitochondrial function are less clear. Despite attenuation of LV and metabolic remodeling [75, 76], ejection fraction and mitochondrial respiration remained unaffected [77, 78] in several animal models of cardiac dysfunction with n-3 PUFA supplementation.

The findings from the above studies would suggest that the ability of dietary strategies to alter the mitochondrial membrane composition in an effort to delay the development of cardiac dysfunction and/or heart failure are not promising. However, careful consideration of several factors must be made. First, some studies report the composition of phospholipids from heart homogenates while others report the composition of phospholipids from isolated mitochondria. It is not clear whether there is a homogenous composition of phospholipids throughout the cardiac myocyte, especially as phospholipid remodeling is a dynamic process that might occur differentially at the sarcolemma versus other organelles [79]. Second, the reported content of unsaturated fatty acids from phospholipids varies significantly between studies. The intricacy of extraction protocols that could render unsaturated fatty acids prone to degradation may be a critical factor in this regard. The variety of mass spectrometry based detection methodologies may also contribute to reported differences. Last, and perhaps most important, is the potential differences in phospholipid fatty acid composition across species [71, 75, 80] that may confound translation of animal studies to human.

Summary and Conclusion

Despite nearly 50 years of research on examining the role of PUFAs and overall health in cell culture systems, animal models, and humans, it remains unclear whether dietary or supplementation strategies are a safe and effective means to improve heart health. Clearly, limiting dietary intake of saturated fatty acids in favor of PUFAs would be a favorable behavioral modification to decrease risk of developing cardiovascular disease in all individuals. This strategy may also be beneficial in patients with cardiovascular disease to reduce mortality and further hospitalizations. However, it is still debatable whether PUFA supplementation, potentially via fish oil, is a worthwhile endeavor. Standardization of clinical protocols, including establishment of a therapeutic index, may help to provide unequivocal conclusions.

Coinciding with the clinical trials, animal models must also adhere to controlled study designs, including treatment duration and dosage, in order to help elucidate the possible molecular underpinnings of n-3 PUFA actions on cardiac function.

REFERENCES

[1] Kolwicz, S. C., Jr., S. Purohit, and R. Tian. (2013). Cardiac metabolism and its interactions with contraction, growth, and survival of cardiomyocytes. *Circ Res*, 113(5): p. 603-16.

[2] Kolwicz, S. C., Jr., L. Liu, I. J. Goldberg, and R. Tian. (2015). Enhancing Cardiac Triacylglycerol Metabolism Improves Recovery From Ischemic Stress. *Diabetes*, 64(8): p. 2817-27.

[3] Kolwicz, S. C., Jr. (2016). Lipid partitioning during cardiac stress. *Biochim Biophys Acta*, 1861(10): p. 1472-80.

[4] Bosma, M., D. H. Dapito, Z. Drosatos-Tampakaki, N. Huiping-Son, L. S. Huang, S. Kersten, K. Drosatos, and I. J. Goldberg. (2014). Sequestration of fatty acids in triglycerides prevents endoplasmic reticulum stress in an in vitro model of cardiomyocyte lipotoxicity. *Biochim Biophys Acta*, 1841(12): p. 1648-55.

[5] Listenberger, L. L., D. S. Ory, and J. E. Schaffer. (2001). Palmitate-induced apoptosis can occur through a ceramide-independent pathway. *J Biol Chem*, 276(18): p. 14890-5.

[6] Russo, S. B., C. F. Baicu, A. Van Laer, T. Geng, H. Kasiganesan, M. R. Zile, and L. A. Cowart. (2012). Ceramide synthase 5 mediates lipid-induced autophagy and hypertrophy in cardiomyocytes. *J Clin Invest*, 122(11): p. 3919-30.

[7] Mozaffarian, D., L. J. Appel, and L. Van Horn. (2011). Components of a cardioprotective diet: new insights. *Circulation*, 123(24): p. 2870-91.

[8] Estruch, R., E. Ros, J. Salas-Salvado, M. I. Covas, D. Corella, F. Aros, E. Gomez-Gracia, V. Ruiz-Gutierrez, M. Fiol, J. Lapetra, R. M. Lamuela-Raventos, L. Serra-Majem, X. Pinto, J. Basora, M. A.

Munoz, J. V. Sorli, J. A. Martinez, M. A. Martinez-Gonzalez, and P. S. Investigators. (2013). Primary prevention of cardiovascular disease with a Mediterranean diet. *N Engl J Med*, 368(14): p. 1279-90.

[9] Esposito, K. and D. Giugliano. (2014). Mediterranean diet and type 2 diabetes. *Diabetes Metab Res Rev*, 30 Suppl 1: p. 34-40.

[10] Schwingshackl, L. and G. Hoffmann. (2016). Does a Mediterranean-Type Diet Reduce Cancer Risk? *Curr Nutr Rep*, 5: p. 9-17.

[11] United States. Department of Health and Human Services., United States. Department of Agriculture., and United States. Dietary Guidelines Advisory Committee., (2015). *Dietary guidelines for Americans, 2015-2020.* Eighth edition. ed. Washington, D. C.: U. S. Department of Health and Human Services and U. S. Department of Agriculture. pgs. 122.

[12] Vannice, G. and H. Rasmussen. (2014). Position of the academy of nutrition and dietetics: dietary fatty acids for healthy adults. *J Acad Nutr Diet*, 114(1): p. 136-53.

[13] Eckel, R. H., J. M. Jakicic, J. D. Ard, J. M. de Jesus, N. Houston Miller, V. S. Hubbard, I. M. Lee, A. H. Lichtenstein, C. M. Loria, B. E. Millen, C. A. Nonas, F. M. Sacks, S. C. Smith, Jr., L. P. Svetkey, T. A. Wadden, S. Z. Yanovski, and G. American College of Cardiology/American Heart Association Task Force on Practice. (2014). 2013 AHA/ACC guideline on lifestyle management to reduce cardiovascular risk: a report of the American College of Cardiology/American Heart Association Task Force on Practice Guidelines. *J Am Coll Cardiol*, 63(25 Pt B): p. 2960-84.

[14] Guillou, H., D. Zadravec, P. G. Martin, and A. Jacobsson. (2010). The key roles of elongases and desaturases in mammalian fatty acid metabolism: Insights from transgenic mice. *Prog Lipid Res*, 49(2): p. 186-99.

[15] Plourde, M. and S. C. Cunnane. (2007). Extremely limited synthesis of long chain polyunsaturates in adults: implications for their dietary essentiality and use as supplements. *Appl Physiol Nutr Metab*, 32(4): p. 619-34.

[16] Bang, H. O., J. Dyerberg, and A. B. Nielsen. (1971). Plasma lipid and lipoprotein pattern in Greenlandic West-coast Eskimos. *Lancet*, 1(7710): p. 1143-5.

[17] Dyerberg, J. and H. O. Bang. (1979). Haemostatic function and platelet polyunsaturated fatty acids in Eskimos. *Lancet*, 2(8140): p. 433-5.

[18] Nodari, S., M. Triggiani, U. Campia, A. Manerba, G. Milesi, B. M. Cesana, M. Gheorghiade, and L. Dei Cas. (2011). Effects of n-3 polyunsaturated fatty acids on left ventricular function and functional capacity in patients with dilated cardiomyopathy. *J Am Coll Cardiol*, 57(7): p. 870-9.

[19] Zhao, Y., S. Joshi-Barve, S. Barve, and L. H. Chen. (2004). Eicosapentaenoic acid prevents LPS-induced TNF-alpha expression by preventing NF-kappaB activation. *J Am Coll Nutr*, 23(1): p. 71-8.

[20] Balk, E. M., A. H. Lichtenstein, M. Chung, B. Kupelnick, P. Chew, and J. Lau. (2006). Effects of omega-3 fatty acids on serum markers of cardiovascular disease risk: a systematic review. *Atherosclerosis*, 189(1): p. 19-30.

[21] Leaf, A., J. X. Kang, Y. F. Xiao, and G. E. Billman. (2003). Clinical prevention of sudden cardiac death by n-3 polyunsaturated fatty acids and mechanism of prevention of arrhythmias by n-3 fish oils. *Circulation*, 107(21): p. 2646-52.

[22] Rasmussen, B. M., B. Vessby, M. Uusitupa, L. Berglund, E. Pedersen, G. Riccardi, A. A. Rivellese, L. Tapsell, K. Hermansen, and K. S. Group. (2006). Effects of dietary saturated, monounsaturated, and n-3 fatty acids on blood pressure in healthy subjects. *Am J Clin Nutr*, 83(2): p. 221-6.

[23] Kris-Etherton, P. M., W. S. Harris, L. J. Appel, and C. American Heart Association. Nutrition. (2002). Fish consumption, fish oil, omega-3 fatty acids, and cardiovascular disease. *Circulation*, 106(21): p. 2747-57.

[24] Galan, P., E. Kesse-Guyot, S. Czernichow, S. Briancon, J. Blacher, S. Hercberg, and S. F. O. C. Group. (2010). Effects of B vitamins and omega 3 fatty acids on cardiovascular diseases: a randomised placebo controlled trial. *BMJ*, 341: p. c6273.

[25] Kohashi, K., A. Nakagomi, Y. Saiki, T. Morisawa, M. Kosugi, Y. Kusama, H. Atarashi, and W. Shimizu. (2014). Effects of eicosapentaenoic acid on the levels of inflammatory markers, cardiac function and long-term prognosis in chronic heart failure patients with dyslipidemia. *J Atheroscler Thromb*, 21(7): p. 712-29.

[26] Kromhout, D., E. J. Giltay, J. M. Geleijnse, and G. Alpha Omega Trial. (2010). n-3 fatty acids and cardiovascular events after myocardial infarction. *N Engl J Med*, 363(21): p. 2015-26.

[27] Rauch, B., R. Schiele, S. Schneider, F. Diller, N. Victor, H. Gohlke, M. Gottwik, G. Steinbeck, U. Del Castillo, R. Sack, H. Worth, H. Katus, W. Spitzer, G. Sabin, J. Senges, and O. S. Group. (2010). OMEGA, a randomized, placebo-controlled trial to test the effect of highly purified omega-3 fatty acids on top of modern guideline-adjusted therapy after myocardial infarction. *Circulation*, 122(21): p. 2152-9.

[28] Yokoyama, M., H. Origasa, M. Matsuzaki, Y. Matsuzawa, Y. Saito, Y. Ishikawa, S. Oikawa, J. Sasaki, H. Hishida, H. Itakura, T. Kita, A. Kitabatake, N. Nakaya, T. Sakata, K. Shimada, K. Shirato, and E. P. A. l .i. s. I. Japan. (2007). Effects of eicosapentaenoic acid on major coronary events in hypercholesterolaemic patients (JELIS): a randomised open-label, blinded endpoint analysis. *Lancet*, 369(9567): p. 1090-8.

[29] Bowen, K. J., W. S. Harris, and P. M. Kris-Etherton. (2016). Omega-3 Fatty Acids and Cardiovascular Disease: Are There Benefits? *Curr Treat Options Cardiovasc Med*, 18(11): p. 69.

[30] O'Connell, T. D., R. C. Block, S. P. Huang, and G. C. Shearer. (2017). omega3-Polyunsaturated fatty acids for heart failure: Effects of dose on efficacy and novel signaling through free fatty acid receptor 4. *J Mol Cell Cardiol*, 103: p. 74-92.

[31] Albert, B. B., J. G. Derraik, D. Cameron-Smith, P. L. Hofman, S. Tumanov, S. G. Villas-Boas, M. L. Garg, and W. S. Cutfield. (2015). Fish oil supplements in New Zealand are highly oxidised and do not meet label content of n-3 PUFA. *Sci Rep*, 5: p. 7928.

[32] Albert, B. B., D. Cameron-Smith, P. L. Hofman, and W. S. Cutfield. (2013). Oxidation of marine omega-3 supplements and human health. *Biomed Res Int*, 2013: p. 464921.

[33] Ito, M. K. (2015). *A Comparative Overview of Prescription Omega-3 Fatty Acid Products*. P T, 40(12): p. 826-57.

[34] Siscovick, D. S., T. A. Barringer, A. M. Fretts, J. H. Wu, A. H. Lichtenstein, R. B. Costello, P. M. Kris-Etherton, T. A. Jacobson, M. B. Engler, H. M. Alger, L. J. Appel, D. Mozaffarian, L. American Heart Association Nutrition Committee of the Council on, H. Cardiometabolic, E. Council on, Prevention, Y. Council on Cardiovascular Disease in the, C. Council on, N. Stroke, and C. Council on Clinical. (2017). Omega-3 Polyunsaturated Fatty Acid (Fish Oil) Supplementation and the Prevention of Clinical Cardiovascular Disease: A Science Advisory from the American Heart Association. *Circulation*, 135(15): p. e867-e884.

[35] Calder, P. C. (2009). Polyunsaturated fatty acids and inflammatory processes: New twists in an old tale. *Biochimie*, 91(6): p. 791-5.

[36] Harris, W. S. and G. C. Shearer. (2014). Omega-6 fatty acids and cardiovascular disease: friend, not foe? *Circulation*, 130(18): p. 1562-4.

[37] Watanabe, S., A. Yoshihisa, Y. Kanno, M. Takiguchi, T. Yokokawa, A. Sato, S. Miura, T. Shimizu, S. Abe, T. Sato, S. Suzuki, M. Oikawa, N. Sakamoto, T. Yamaki, K. Sugimoto, H. Kunii, K. Nakazato, H. Suzuki, S. I. Saitoh, and Y. Takeishi. (2016). Associations with Eicosapentaenoic Acid to Arachidonic Acid Ratio and Mortality in Hospitalized Heart Failure Patients. *J Card Fail*, 22(12): p. 962-969.

[38] Tavazzi, L., A. P. Maggioni, R. Marchioli, S. Barlera, M. G. Franzosi, R. Latini, D. Lucci, G. L. Nicolosi, M. Porcu, G. Tognoni, and H. F. I. Gissi. (2008). Effect of n-3 polyunsaturated fatty acids in

patients with chronic heart failure (the GISSI-HF trial): a randomised, double-blind, placebo-controlled trial. *Lancet*, 372(9645): p. 1223-30.

[39] Moertl, D., A. Hammer, S. Steiner, R. Hutuleac, K. Vonbank, and R. Berger. (2011). Dose-dependent effects of omega-3-polyunsaturated fatty acids on systolic left ventricular function, endothelial function, and markers of inflammation in chronic heart failure of nonischemic origin: a double-blind, placebo-controlled, 3-arm study. *Am Heart J*, 161(5): p. 915 e1-9.

[40] Chrysohoou, C., G. Metallinos, G. Georgiopoulos, D. Mendrinos, A. Papanikolaou, N. Magkas, C. Pitsavos, G. Vyssoulis, C. Stefanadis, and D. Tousoulis. (2016). Short term omega-3 polyunsaturated fatty acid supplementation induces favorable changes in right ventricle function and diastolic filling pressure in patients with chronic heart failure; A randomized clinical trial. *Vascul Pharmacol*, 79: p. 43-50.

[41] Burr, M. L., P. A. Ashfield-Watt, F. D. Dunstan, A. M. Fehily, P. Breay, T. Ashton, P. C. Zotos, N. A. Haboubi, and P. C. Elwood. (2003). Lack of benefit of dietary advice to men with angina: results of a controlled trial. *Eur J Clin Nutr*, 57(2): p. 193-200.

[42] Yndestad, A., J. K. Damas, E. Oie, T. Ueland, L. Gullestad, and P. Aukrust. (2006). Systemic inflammation in heart failure--the whys and wherefores. *Heart Fail Rev*, 11(1): p. 83-92.

[43] Liu, L., Y. Wang, Z. Y. Cao, M. M. Wang, X. M. Liu, T. Gao, Q. K. Hu, W. J. Yuan, and L. Lin. (2015). Up-regulated TLR4 in cardiomyocytes exacerbates heart failure after long-term myocardial infarction. *J Cell Mol Med*, 19(12): p. 2728-40.

[44] Balakumar, P. and M. Singh. (2006). Anti-tumour necrosis factor-alpha therapy in heart failure: future directions. *Basic Clin Pharmacol Toxicol*, 99(6): p. 391-7.

[45] Maier, H. J., T. G. Schips, A. Wietelmann, M. Kruger, C. Brunner, M. Sauter, K. Klingel, T. Bottger, T. Braun, and T. Wirth. (2012). Cardiomyocyte-specific IkappaB kinase (IKK)/NF-kappaB activation induces reversible inflammatory cardiomyopathy and heart failure. *Proc Natl Acad Sci USA*, 109(29): p. 11794-9.

[46] Nguyen, S., D. Shao, L. C. Tomasi, A. Braun, A. B. M. de Mattos, Y. S. Choi, O. Villet, N. Roe, C. R. Halterman, R. Tian, and S. C. Kolwicz, Jr. (2017). The effects of fatty acid composition on cardiac hypertrophy and function in mouse models of diet-induced obesity. *J Nutr Biochem*, 46: p. 137-142.

[47] Farvid, M. S., M. Ding, A. Pan, Q. Sun, S. E. Chiuve, L. M. Steffen, W. C. Willett, and F. B. Hu. (2014). Dietary linoleic acid and risk of coronary heart disease: a systematic review and meta-analysis of prospective cohort studies. *Circulation*, 130(18): p. 1568-78.

[48] Johnson, G. H. and K. Fritsche. (2012). Effect of dietary linoleic acid on markers of inflammation in healthy persons: a systematic review of randomized controlled trials. *J Acad Nutr Diet*, 112(7): p. 1029-41, 1041 e1-15.

[49] Grygiel-Gorniak, B. (2014). Peroxisome proliferator-activated receptors and their ligands: nutritional and clinical implications--a review. *Nutr J*, 13: p. 17.

[50] Lo Verme, J., J. Fu, G. Astarita, G. La Rana, R. Russo, A. Calignano, and D. Piomelli. (2005). The nuclear receptor peroxisome proliferator-activated receptor-alpha mediates the anti-inflammatory actions of palmitoylethanolamide. *Mol Pharmacol*, 67(1): p. 15-9.

[51] Suthahar, N., W. C. Meijers, H. H. W. Sillje, and R. A. de Boer. (2017). From Inflammation to Fibrosis-Molecular and Cellular Mechanisms of Myocardial Tissue Remodelling and Perspectives on Differential Treatment Opportunities. *Curr Heart Fail Rep*, 14(4): p. 235-250.

[52] Endo, J. and M. Arita. (2016). Cardioprotective mechanism of omega-3 polyunsaturated fatty acids. *J Cardiol*, 67(1): p. 22-7.

[53] Chen, J., G. C. Shearer, Q. Chen, C. L. Healy, A. J. Beyer, V. B. Nareddy, A. M. Gerdes, W. S. Harris, T. D. O'Connell, and D. Wang. (2011). Omega-3 fatty acids prevent pressure overload-induced cardiac fibrosis through activation of cyclic GMP/protein kinase G signaling in cardiac fibroblasts. *Circulation*, 123(6): p. 584-93.

[54] Harris, W. S. (2010). The omega-3 index: clinical utility for therapeutic intervention. *Curr Cardiol Rep*, 12(6): p. 503-8.

[55] Eclov, J. A., Q. Qian, R. Redetzke, Q. Chen, S. C. Wu, C. L. Healy, S. B. Ortmeier, E. Harmon, G. C. Shearer, and T. D. O'Connell. (2015). EPA, not DHA, prevents fibrosis in pressure overload-induced heart failure: potential role of free fatty acid receptor 4. *J Lipid Res*, 56(12): p. 2297-308.

[56] Fukunaga, S., S. Setoguchi, A. Hirasawa, and G. Tsujimoto. (2006). Monitoring ligand-mediated internalization of G protein-coupled receptor as a novel pharmacological approach. *Life Sci*, 80(1): p. 17-23.

[57] Marchioli, R., F. Barzi, E. Bomba, C. Chieffo, D. Di Gregorio, R. Di Mascio, M. G. Franzosi, E. Geraci, G. Levantesi, A. P. Maggioni, L. Mantini, R. M. Marfisi, G. Mastrogiuseppe, N. Mininni, G. L. Nicolosi, M. Santini, C. Schweiger, L. Tavazzi, G. Tognoni, C. Tucci, F. Valagussa, and G. I.-P. Investigators. (2002). Early protection against sudden death by n-3 polyunsaturated fatty acids after myocardial infarction: time-course analysis of the results of the Gruppo Italiano per lo Studio della Sopravvivenza nell'Infarto Miocardico (GISSI)-Prevenzione. *Circulation*, 105(16): p. 1897-903.

[58] Leaf, A., Y. F. Xiao, J. X. Kang, and G. E. Billman. (2005). Membrane effects of the n-3 fish oil fatty acids, which prevent fatal ventricular arrhythmias. *J Membr Biol*, 206(2): p. 129-39.

[59] Xiao, Y. F., L. Ma, S. Y. Wang, M. E. Josephson, G. K. Wang, J. P. Morgan, and A. Leaf. (2006). Potent block of inactivation-deficient Na+ channels by n-3 polyunsaturated fatty acids. *Am J Physiol Cell Physiol*, 290(2): p. C362-70.

[60] Xiao, Y. F., D. C. Sigg, and A. Leaf. (2005). The antiarrhythmic effect of n-3 polyunsaturated fatty acids: modulation of cardiac ion channels as a potential mechanism. *J Membr Biol*, 206(2): p. 141-54.

[61] Xiao, Y. F., D. C. Sigg, M. R. Ujhelyi, J. J. Wilhelm, E. S. Richardson, and P. A. Iaizzo. (2008). Pericardial delivery of omega-3 fatty acid: a novel approach to reducing myocardial infarct sizes and arrhythmias. *Am J Physiol Heart Circ Physiol*, 294(5): p. H2212-8.

[62] Pound, E. M., J. X. Kang, and A. Leaf. (2001). Partitioning of polyunsaturated fatty acids, which prevent cardiac arrhythmias, into phospholipid cell membranes. *J Lipid Res*, 42(3): p. 346-51.

[63] Mozaffarian, D., R. Marchioli, A. Macchia, M. G. Silletta, P. Ferrazzi, T. J. Gardner, R. Latini, P. Libby, F. Lombardi, P. T. O'Gara, R. L. Page, L. Tavazzi, G. Tognoni, and O. Investigators. (2012). Fish oil and postoperative atrial fibrillation: the Omega-3 Fatty Acids for Prevention of Post-operative Atrial Fibrillation (OPERA) randomized trial. *JAMA*, 308(19): p. 2001-11.

[64] Sandesara, C. M., M. K. Chung, D. R. Van Wagoner, T. A. Barringer, K. Allen, H. M. Ismail, B. Zimmerman, and B. Olshansky. (2012). A Randomized, Placebo-Controlled Trial of Omega-3 Fatty Acids for Inhibition of Supraventricular Arrhythmias After Cardiac Surgery: The FISH Trial. *J Am Heart Assoc*, 1(3): p. e000547.

[65] Billman, G. E., C. A. Carnes, P. B. Adamson, E. Vanoli, and P. J. Schwartz. (2012). Dietary omega-3 fatty acids and susceptibility to ventricular fibrillation: lack of protection and a proarrhythmic effect. *Circ Arrhythm Electrophysiol*, 5(3): p. 553-60.

[66] Coronel, R., F. J. Wilms-Schopman, H. M. Den Ruijter, C. N. Belterman, C. A. Schumacher, T. Opthof, R. Hovenier, A. G. Lemmens, A. H. Terpstra, M. B. Katan, and P. Zock. (2007). Dietary n-3 fatty acids promote arrhythmias during acute regional myocardial ischemia in isolated pig hearts. *Cardiovasc Res*, 73(2): p. 386-94.

[67] Chicco, A. J. and G. C. Sparagna. (2007). Role of cardiolipin alterations in mitochondrial dysfunction and disease. *Am J Physiol Cell Physiol*, 292(1): p. C33-44.

[68] Osman, C., D. R. Voelker, and T. Langer. (2011). Making heads or tails of phospholipids in mitochondria. *J Cell Biol*, 192(1): p. 7-16.

[69] Horvath, S. E. and G. Daum. (2013). Lipids of mitochondria. *Prog Lipid Res*, 52(4): p. 590-614.

[70] Mejia, E. M. and G. M. Hatch. (2015). Mitochondrial phospholipids: role in mitochondrial function. *J Bioenerg Biomembr*.

[71] Le, C. H., C. M. Mulligan, M. A. Routh, G. J. Bouma, M. A. Frye, K. M. Jeckel, G. C. Sparagna, J. M. Lynch, R. L. Moore, S. A. McCune, M. Bristow, S. Zarini, R. C. Murphy, and A. J. Chicco. (2014). Delta-6-desaturase links polyunsaturated fatty acid metabolism with phospholipid remodeling and disease progression in heart failure. *Circ Heart Fail*, 7(1): p. 172-83.

[72] Lee, H. J., J. Mayette, S. I. Rapoport, and R. P. Bazinet. (2006). Selective remodeling of cardiolipin fatty acids in the aged rat heart. *Lipids Health Dis*, 5: p. 2.

[73] Nasa, Y., Y. Sakamoto, A. Sanbe, H. Sasaki, F. Yamaguchi, and S. Takeo. (1997). Changes in fatty acid compositions of myocardial lipids in rats with heart failure following myocardial infarction. *Mol Cell Biochem*, 176(1-2): p. 179-89.

[74] Sparagna, G. C., A. J. Chicco, R. C. Murphy, M. R. Bristow, C. A. Johnson, M. L. Rees, M. L. Maxey, S. A. McCune, and R. L. Moore. (2007). Loss of cardiac tetralinoleoyl cardiolipin in human and experimental heart failure. *J Lipid Res*, 48(7): p. 1559-70.

[75] O'Shea, K. M., D. J. Chess, R. J. Khairallah, P. A. Hecker, B. Lei, K. Walsh, C. Des Rosiers, and W. C. Stanley. (2010). omega-3 Polyunsaturated fatty acids prevent pressure overload-induced ventricular dilation and decrease in mitochondrial enzymes despite no change in adiponectin. *Lipids Health Dis*, 9: p. 95.

[76] Shah, K. B., M. K. Duda, K. M. O'Shea, G. C. Sparagna, D. J. Chess, R. J. Khairallah, I. Robillard-Frayne, W. Xu, R. C. Murphy, C. Des Rosiers, and W. C. Stanley. (2009). The cardioprotective effects of fish oil during pressure overload are blocked by high fat intake: role of cardiac phospholipid remodeling. *Hypertension*, 54(3): p. 605-11.

[77] Galvao, T. F., R. J. Khairallah, E. R. Dabkowski, B. H. Brown, P. A. Hecker, K. A. O'Connell, K. M. O'Shea, H. N. Sabbah, S. Rastogi, C. Daneault, C. Des Rosiers, and W. C. Stanley. (2013). Marine n3 polyunsaturated fatty acids enhance resistance to mitochondrial permeability transition in heart failure but do not improve survival. *Am J Physiol Heart Circ Physiol*, 304(1): p. H12-21.

[78] O'Shea, K. M., R. J. Khairallah, G. C. Sparagna, W. Xu, P. A. Hecker, I. Robillard-Frayne, C. Des Rosiers, T. Kristian, R. C. Murphy, G. Fiskum, and W. C. Stanley. (2009). Dietary omega-3 fatty acids alter cardiac mitochondrial phospholipid composition and delay $Ca2^{+}$-induced permeability transition. *J Mol Cell Cardiol*, 47(6): p. 819-27.

[79] Hishikawa, D., T. Hashidate, T. Shimizu, and H. Shindou. (2014). Diversity and function of membrane glycerophospholipids generated by the remodeling pathway in mammalian cells. *J Lipid Res*, 55(5): p. 799-807.

[80] Harris, W. S., S. A. Sands, S. L. Windsor, H. A. Ali, T. L. Stevens, A. Magalski, C. B. Porter, and A. M. Borkon. (2004). Omega-3 fatty acids in cardiac biopsies from heart transplantation patients: correlation with erythrocytes and response to supplementation. *Circulation*, 110(12): p. 1645-9.

In: Polyunsaturated Fatty Acids (PUFAs) ISBN: 978-1-53613-572-5
Editor: Angel Catala

Chapter 7

PLANTS AS ALTERNATIVE SOURCES OF N-3 POLYUNSATURATED FATTY ACIDS

Javier S. Perona*, Silvia Garcia-Rodriguez and Jose M. Castellano
Department of Food and Health,
Instituto de la Grasa-CSIC, Seville, Spain

ABSTRACT

It is well accepted that the consumption of very long-chain n-3 PUFA of marine origin have beneficial effects on plasma triglyceride concentrations and on systemic inflammation states. α-Linolenic acid (18: 3 n-3, ALA), although also a member of the n-3 PUFA family, is found predominantly in plants and its beneficial effects on health are more controversial. However, it has been suggested that ALA intake may reduce the risk of coronary heart disease when the intake of n-3 PUFA from marine sources is low. It is still unclear whether the effect is per se, or after its bioconversion to eicosapentaenoic (EPA) or docosahexaenoic (DHA) n-3 PUFA. Despite the recommendations of international institutions, including the World Health Organization, to increase fish

* Corresponding Author: perona@ig.csic.es.

consumption, in many countries it is difficult to achieve sufficient intake of n-3 PUFA. Access to n-3 PUFA of marine origin may be limited even in those regions with access to the sea. On the other hand, fish stocks are endangered around the world, making aquaculture the only sustainable solution for the future. Thus, the consumption of ALA-rich plant food is postulated as one of the most realistic alternatives to endure the n-3 PUFA intake. Among the different alternatives of plant n-3 PUFA-rich sources, chia (*Salvia hispanica*), sacha inchi (*Plukenetia volubilis*) and linseed (*Linum usitatissimum)*, have been considered. Oils obtained from their seeds can provide up to 65% of ALA. In addition, nuts are also appreciated sources of n-3 PUFA. Finally, transgenic plants have also been proposed as sustainable sources of EPA and DHA. We review here the current knowledge on plant n-3 PUFA consumption and the physiological mechanisms, mediated by triglyceride synthesis and the inhibition of inflammatory processes, by which they can exert beneficial effects on human health.

1. Introduction

Few dietary components are so widely recognized as able to improve the human health as the n-3 PUFA. The shelves of supermarkets and food stores exhibit many products with labels indicating the presence of these compounds. The prevalent PUFA found in nature belongs to the n-3 and n-6 classes, where the double bond that is closest to the methyl terminus of the acyl chain is located between carbons 3 and 4, and between carbons 6 and 7, respectively. The three main dietary n-3 PUFA are α-linolenic acid (18:3 n-3, all-cis-octadeca-9,12,15-trienoic acid, ALA), eicosapentaenoic acid (20:5 n-3, all-cis-eicosa-5,8,11,14,17-pentaenoic acid, EPA), and docosahexaenoic acid (22:6 n-3, all-cis-docosa-4,7,10,13,16,19-hexaenoic acid, DHA). Plants can desaturate fatty acids at positions 3, 6 and 9, while animals, particularly vertebrates (including mammals), can only introduce unsaturations from carbon 9 towards the carboxyl group. For this reason, linoleic acid (18:2 n-6, all-cis-octadeca-9,12-dienoic acid, LA) and ALA can not be synthesized in mammals from precursors of lower unsaturation, and must be provided by the diet in certain quantities and proportions. Thus, LA and ALA are recognized as essential fatty acids.

Although animals cannot synthesize ALA, they can metabolize it to other longer chain and more unsaturated n-3 PUFA. This occurs by a series of linked desaturation and elongation reactions that mainly take place in the liver. Firstly, ALA is converted to stearidonic acid (18:4 n-3) by Δ6-desaturase, and then elongated to eicosatetraenoic acid (20:4 n-3). Further desaturation by Δ5-desaturase yields EPA. Conversion of ALA to EPA competes with the biosynthesis of arachidonic acid (AA; 20:4 n-6) from LA, since both metabolic pathways use the same microsomal enzyme system. EPA can be further converted to DHA through a four-step process. Initially, EPA is elongated to form docosapentaenoic acid (22:5 n-3, DPA), which is subsequently elongated and unsaturated by Δ6-desaturase to produce 24:6 n-3. This fatty acid is then transported to the peroxisome, where it is converted in DHA by partial β-oxidation.

It is assumed that Δ6-desaturase is rate limiting in this pathway. The activities of Δ6-and Δ5-desaturases are regulated by nutritional status, hormones, and feedback inhibition by end products, creating a complex control network for endogenous long-chain PUFA synthesis.

Once consumed in the diet, n-3 PUFA are absorbed from the gastrointestinal tract and transport to the liver via chylomicron particles. These fatty acids occupy, preferentially, the sn-2 position in phospholipids and triglycerides, so they are not released by the intestinal pancreatic lipase (with specificity for sn-1 and sn-3), being absorbed as lysophospholipids or monoglycerides. Nor are they released from the circulating chylomicrons by the vascular lipoprotein lipase. In the liver, n-3 PUFA are used as a source of triglycerides in lipoprotein particles and phospholipids. From the liver, n-3 PUFA are released into the blood mainly as phospholipids associated to plasma albumin, and incorporated into cell membrane phospholipids throughout the body. In addition, some of them are stored in the adipose tissue as triglycerides. The conversion of 24:6 n-3 in DHA would occur, above all, in the astrocytes of the glia. These cells have the function of providing DHA to the neurons, since this fatty acid is essential for maintaining the fluidity of the plasma membrane and for the formation of other bioactive compounds.

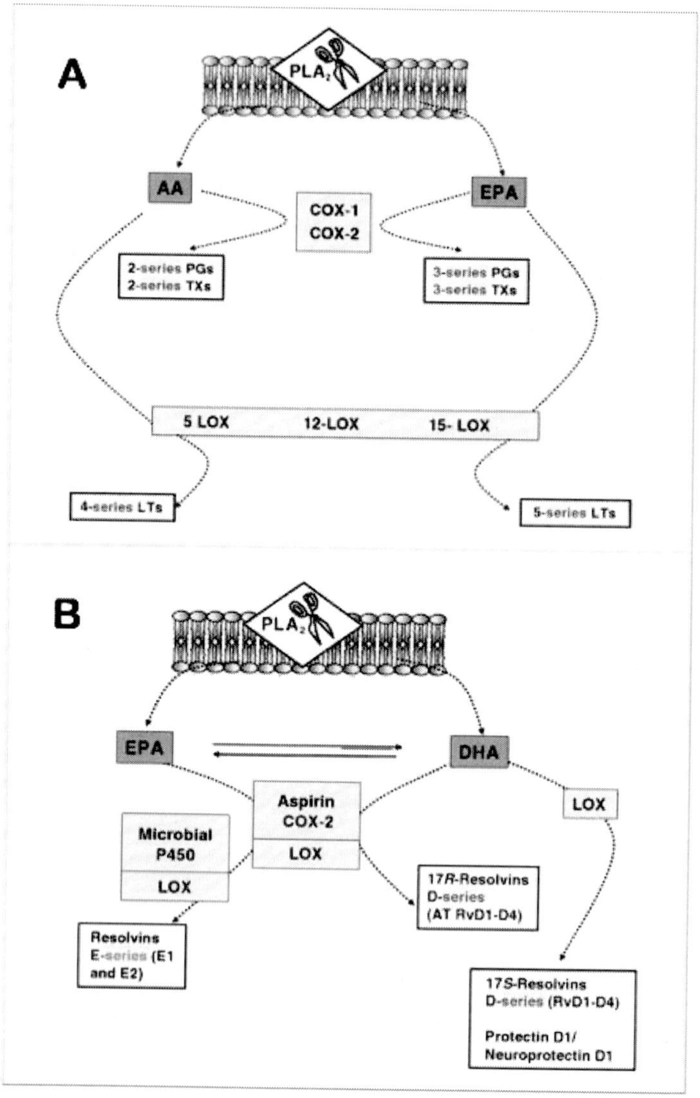

Figure 1. Synthesis of eicosanoids and docosanoids. A) Competition between arachidonic acid (AA) and eicosapentaenoic acid (EPA) for cyclooxygenases (COX) and lypoxygenases (LOX). After be released from membrane phospholipids by the action of phospholipase A2, free AA and EPA are converted by the same enzymes (COX and LOX) to their oxygenated metabolites prostaglandins (PGs), tromboxanes (TXs) and leukotrienes (LTs), collectively named eicosanoids. B) Formation of resolvins and protectins from EPA and docosahexaenoic acid (DPA). (Adapted from G. Calviello and S. Serini. Dietary Omega-3 Polyunsaturated Fatty Acids and Cancer. Springer Science + Business Media B. V. 2010. ISBN 978-90-481-3578-3).

The sn-2 position of membrane glycerophospholipids is devoted to AA, an n-6 PUFA. AA is released from membrane phospholipids by phospholipase A2, an enzyme that can be activated by inflammation. The free AA is then processed by several enzymes belonging to the COX and LOX families as well as cytochrome P450, to generate potent pro-inflammatory eicosanoids (Lewis et al., 1990; Tilley et al., 2001; Kroetz and Zeldin, 2002) (Figure 1).

Since n-3 and n-6 PUFA are metabolized by the same enzymes, COXs and LOXs, n-3 PUFA compete with n-6 PUFA for these enzymes and inhibit biosynthesis of n-6 series eicosanoid (Berquin et al., 2008). Several research groups reported that n-3 PUFA counter-regulate AA-derived eicosanoids in cells, animals, and humans by inhibiting n-6 PUFA metabolism and antagonizing them on their oxygenation pathways to produce mediators (Edwards and O'Flaherty, 2008; Gleissman et al., 2010; Greene et al., 2011). DHA can downregulate the formation of AA-derived PGE2. Thus, dietary n-3 PUFA may function as natural COX inhibitors.

EPA and DHA are also released intracellularly from membranes by the action of phospholipase A2, and subsequently metabolized by the cyclooxygenase (COX) and lipoxygenase (LOX) enzymes. The action of COX on EPA produces 3-series thromboxanes (TX_3), prostaglandins (PG_3) and prostacyclines (PCI_3). Likewise, LOX transforms EPA into 5-series leukotrienes (LT_5). These EPA metabolites have, generally, little biological activity, but they interfere with the action of analog eicosanoids synthesized from the AA cascade, generally pro-inflammatory. In addition, it has been reported that the PCI_3 formed in the endothelial cells exert inhibitory effects on platelet aggregation and are vasodilators. Likewise, LT_5 originated in leukocytes have anti-inflammatory effects and inhibit chemotaxis and cell adhesion.

1.1. Dietary Sources and Bioavailability of n-3 PUFA

The major dietary source of EPA and DHA are fish, especially those from cold water. It is well known that fish fat contains compounds of dietary and pharmacological interest, and that their physical properties make them useful as emollients. Two types of fish can be distinguished: a first category composed of fish such as cod or haddock, whose primary reserve of lipids is the liver (these fish are often known as white fish or lean, with <2% fat), and a second category that includes fish whose fat is distributed in its flesh (muscle), which gives it a whitish tone, which is sometimes colored by the presence of carotenoids such as astaxanthin. These fish are known as fatty or blue fish, with >5% fat, and sardines, anchovies, herring or salmon are typical examples. The content of EPA and DHA of the different marine species varies depending on the season, whether they are caught in the open sea or comes from fish farming and the method of cooking. EPA and DHA may be also artificially added to the diet as fish oil supplements, in capsule format or as fish oil-enriched food. In this sense, the fishing industry produces flours and oils from fishing waste or from species caught for this purpose. Currently, fish oil capsules are commercialized, which generally contain around 18% of EPA and 12% of DHA, as well as products based on sardine and anchovy oils, from particular regions, captured at specific times, presenting an EPA: DHA ratio of 3-4: 1.

Nowadays, both PUFA n-6 and n-3 constitute a good model for the development of functional foods and nutraceuticals. Efforts have been made to incorporate them into milk drinks and derivatives. Margarines, shortenings and mayonnaises are made enriched in ALA, EPA and DHA, what is an adequate way to supplement the diet of populations in which fish consumption is reduced. Bioavailability of n-3 PUFA in humans has been evaluated by measuring their concentrations in plasma, blood cells and different tissues. It has been reported that a dietary fish oil supplementation (3 g/day EPA + DHA) or a daily serving of salmon (1.2 g/day EPA + DHA) may allow an enrichment of EPA and DHA in serum lipids, ranging from 100 to 130% for EPA, and from 25 to 40% for DHA. Even higher increases were reported for total phospholipids after dietary supplementation with EPA and DHA ethyl esters (1.9 and 0.9 g/day, respectively). It is not definitely established what is the best source for maximal bioavailability of these fatty acids. However, it has been reported that high plasma concentrations of EPA and DHA was better achieved if their source was fish (containing mainly n-3 PUFA esterified as acylglycerols), rather than capsules containing the fatty acids ethyl ester derivatives.

Thus, the latest consensus recommendations and practice guidelines advice a regular consumption of fish (two or three servings a week), especially of the small fatty species (sardine, anchovy, mackerel, herring). Each serving, on average, should provide 200-500 mg of EPA + DHA. The dietary intake of EPA and DHA has been also recommended during pregnancy and lactation, since it is well established that these fatty acids exert crucial effects on growth and neurological development of fetuses and newborns infants. The maternal plasma phospholipid concentration of PUFA increases during pregnancy, probably mobilized from maternal stores. It was suggested that DHA could be stored in the mother's adipose tissue, as temporary reservoir. It is recommended that the mother receive 300 mg/day of EPA + DHA, of which, the consumption of DHA should correspond to 200 mg/day. However, intakes of up to 2.7 g of EPA + DHA per day (with 1 g/day of DHA) are accepted as the upper limit without adverse reactions (FAO, 2010). Newborns who receive breastfeeding adequately meet their n-3 PUFA requirements, since breast milk, on average, contributes 0.2-0-4% of DHA.

1.2. Beneficial Effects of n-3 PUFA on Human Health

The role of n-3 PUFA as nutritional factors with potential to prevent or delay the progression of prevalent chronic pathologies has been established. It is recognized that n-3 PUFA improve dyslipidemia, especially lowering plasma levels of triglycerides, slightly decrease blood pressure, inhibit the formation of the atheroma plaque, and reduce the risk of sudden death, arrhythmia and stroke, in individuals diagnosed of heart pathologies. Furthermore, these fatty acids can be useful in preventing the vascular complications of diabetes.

The importance of an increased dietary intake of n-3 PUFA has been also recognized for the prevention of neurodegenerative processes. Epidemiological studies have indicated the possibility that dietary EPA and DHA may modify the risk and progression of the Alzheimer's disease. Similarly, the benefits of EPA and DHA have been demonstrated in immunity and inflammatory disorders.

1.2.1. Effects of n-3 PUFA in Cardiovascular Diseases

Certain countries, such as those in the Mediterranean basin and Japan, are known to have a surprisingly low rate of coronary heart disease (CHD) and cardiac death. This could be partly explained by regional dietary factors, as suggested by a common feature of the diet of these countries: the high consumption of fish, the main source of long chain n-3 PUFA. The number of epidemiological studies and controlled clinical trials that have evaluated the effects of EPA and DHA (or their food sources) in relation to cardiovascular risk factors and clinical events, is greater than for any other food or nutrient.

However, in the last years, a number of meta-analyses of epidemiological observations and randomized clinical trials focused on n-3 PUFA have not obtained conclusive results with respect to CHD, although they have detected a consistent protection against cardiac mortality. A possible explanation of this could be purely methodological, due to the variability in the selection of the included studies, the variability in the inclusion/exclusion criteria, the stated objectives, the type and duration of the interventions, the verification of adherence to the same or the statistical analysis itself.

The null results could also be explained by other reasons, such as the substantial improvement in the guidelines for the treatment of ischemic patients in the last years, or the high basal consumption of fish, EPA and DHA or fish oil capsules, by patients who are aware of having a high cardiovascular risk, that are also well informed of the benefits attached to these PUFA. Under these conditions, the 1 g/day increase in EPA + DHA consumption could hardly demonstrate efficacy in reducing cardiovascular risk. n-3 PUFA have pleiotropic effects in addition to lipid modifying effects. Both pre-clinical and clinical studies have shown that n-3 PUFA prevent cardiovascular events by ameliorating endothelial function and attenuating lipid accumulation, vascular inflammation and macrophage recruitment.

1.2.2. Effects of n-3 PUFA in Insulin Resistance and Diabetes

There is not a clear consensus from human trials on the systematic use of n-3 PUFA supplements for people with insulin resistance or type 2 diabetes. Substantial inconsistencies exist between studies of humans compared to rodent models. While most studies in rodents suggest a favorable effect of n-3 fatty acids on glucose homeostasis and insulin sensitivity, human studies are conflicting. Although some report improvement in insulin sensitivity by fish oil consumption, the majority of studies do not corroborate the findings.

Most of observational studies reflect a beneficial effect of n-3 PUFA on insulin action, describing an inverse association between fatty acid level and the homeostasis model assessment of insulin resistance (HOMA-IR) index, in populations with a high intake of fish and a low risk for metabolic syndrome and diabetes type 2, indicating that n-3 PUFA could assist in the prevention and treatment of insulin resistance in humans (Connor et al., 2007; Serhan et al., 2008; Gleissman et al., 2010). This association remained significant even after adjusting for age, gender, BMI, and ethnicity (Connor et al., 2007). Nevertheless, in a prospective study in Korean healthy adults, who were followed for three years, the consumption of fish oil was associated with a lower risk for metabolic syndrome among men, but not among women (Oh et al., 2010).

In contrast to observational studies, in most of the identified randomized controlled trials there was no appreciated changes in insulin sensitivity. A systematic meta-analysis of 11 randomized clinical trials published until 2010, with 618 participants, concluded that n-3 PUFA consumption did not affect insulin sensitivity (Teng et al., 2012). Another systematic review, which examined the effect of EPA and DHA on metabolic syndrome risk factors, concluded that there are no clear effects on selected markers, except for an improvement in blood pressure and the well-established hypotriglyceridemic effect. Surprisingly, lower doses of n-3 PUFA were associated with further benefits of reducing pro-atherogenic low density lipoproteins, whereas greater doses (>3 g/day) were associated with increases in LDL cholesterol (Hu et al., 2010).

It is well assumed that n-3 PUFA do not provide beneficial effects on the glycemic control of patients with established type 2 diabetes. Two meta-analyses, which included 18 and 23 randomized trials, respectively, involving a large numbers of participants, were concordant in outcomes, describing that n-3 PUFA supplementation cause a decrease in plasma triglycerides, a slight increase in LDL, but no effect on the glycemic control or fasting insulin (Dyerberg et al., 1975; Jatoi et al., 2004).

1.2.3. Effects of n-3 PUFA in Cancer

There is increasing evidence that n-3 PUFA play a key role in cellular homeostasis. From this idea has arisen the hypothesis that alterations in the intake and/or metabolism of these fatty acids can alter cellular functions, thus modifying the progression of tumor cells. In this sense, the n-3 PUFA have demonstrated antitumor properties both in vitro and in animal models of cancer (Azrad et al., 2013). However, the evidence in humans is not so clear and no significant associations have been found between the paradigmatic cancers of Western society (breast, prostate and colorectal) and the intake of n-3 PUFA (Gerber, 2012) or the consumption of fish (Sala-Vila and Calder, 2011). Population-based studies mostly rely on data from self-reported dietary fat consumption or from assessments based on national dietary habits, and these evaluations can be poorly correlated to real fatty acid composition in patient samples by direct measurements. In addition, the actual amount of n-3 PUFA consumption may be too low to have a protective effect in some cases. Similarly, the ratio of n-6/n-3 PUFA could be more important than the absolute amount of n-3 PUFA, as suggested by animal and human studies (Gu et al., 2013).

Traditional cancer chemotherapy often does not provide satisfactory long-term clinical results. In most cases, only partial response is achieved, and cancer cells often continue to proliferate and eventually metastasize. Combining agents result in superior response rates and increased disease-free or even overall survival. However, combinational chemotherapy has also been associated with increased treatment complexity and toxicity and, frequently, decreased quality of life (Huober and Thurlimann, 2009). n-3 PUFA are good candidates for chemoprevention or combinational chemotherapy, as health promoters to increase the quality of life of cancer patients. It has been reported that plasma n-3 PUFA levels in cancer patients are up to 50% lower than in healthy individuals, and this is associated to loss of adipose tissue and skeletal muscle. This suggests that supplementation with n-3 PUFA may be beneficial. Many studies show prevention of muscle loss or gain of body mass with n-3 PUFA supplementation during cancer chemotherapy. n-3 PUFA supplementation may not only improve the cachexia condition for cancer patients but also deliver better response to treatment and reduce side effects associated with cancer chemotherapy (Murphy et al., 2012). Many clinical intervention trials have been proposed and developed to validate the effectiveness of n-3 PUFA in cancer prevention and treatment or to provide nutritional support for cancer patient who suffered weight loss, fatigue, and other inflammation-related illness (Berquin et al., 2008).

The inflammatory reaction to a local tumor can also trigger a cascade of systemic inflammation that eventually lead to development of anorexia and catabolic processes, such as muscle proteolysis and lipolysis, the early stage of fatigue and cachexia (Gullett et al., 2011). Several clinical trials, using EPA or marine fish oil (EPA + DHA) in purified form or in the form of oral supplements, have showed that EPA supplementation (>2 g/day) can reduce, even stabilize, losses of weight and muscle mass in patients at advanced stages of cancer (Barberet al., 1999; Fearon et al., 2003; Colomer et al., 2007; Gullett et al., 2011). However, data from other controlled randomized trials did not report benefits by the use of EPA (Jatoi et al., 2004; Fearon et al., 2006). Successful management of cachexia may require a multimodal approach with nutritional supplementation and pharmacological treatment. A recent randomized phase III clinical trial suggested that combination therapy with EPA supplementation, megestrol acetate, thalidomide, and L-carnitine was significantly more effective in improving lean body mass and appetite than single agents (Mantovani,

2010). It is believed that tumor cells can be more easily sensitized to chemotherapy when membrane lipids contain high level of DHA. A clinical phase II trial to improve outcome of anthracycline chemotherapy by using DHA (1.8 g/day) in metastatic breast cancer patients has demonstrated the delay of tumor progression and a higher overall survival. The study also found that DHA had no adverse side effects, unlike chemotherapy (Bougnoux et al., 2009).

2. n-3 PUFA of Vegetable Origin. Seeds Rich in Alpha-Linolenic Acid

As earlier mentioned, ALA is an essential fatty acid that can be elongated and desaturated to become very long-chain PUFA, such as EPA and DHA (Morales et al., 2012). Despite the recommendations of the World Health Organization, the intake of ALA has been reduced in Western societies. One of the main contributors to the reduction of ALA consumption was the increase in the intake of grains with a high content of LA, such as corn. In Western Europe and North America, this has drastically changed the proportion of n-6:n-3 fatty acids from 8:1 to 20:1, which is far from what is considered optimal (4:1) (Barceló-Coblijn et al., 2003).

Unfortunately, it is not an easy goal to increase the intake of n-3 PUFA due to the scarcity of natural sources and their limited availability. The natural sources that contain high levels of EPA and DHA are basically limited to fatty fish and to a number of other types of seafood, which provides sustainability problems when it comes to encouraging the population to increase their consumption (Barceló-Coblijn et al., 2009). Therefore, a suggested alternative would be the vegetable sources of ALA, particularly in countries with little or no access to fish and seafood. Variable amounts of ALA are found in plants, zooplankton, phytoplankton and marine species. In plants, ALA is found in leaves, mainly in glycolipids, and as triglycerides in certain seed oils, beans and nuts.

Table 1. Alpha-linolenic acid (ALA) content in ALA-rich vegetable sources

Common name	Scientific name	g ALA/100g
Flaxseed	*Linum usitatissimum*	22.8
Flaxseed oil		53.3
Perilla	*Perilla frutescens*	58
Perilla oil		55
Chia seed	*Salvia hispanica*	17.6
Chia oil		65
Sclarea seed	*Salvia sclarea*	15
Sclarea oil		60
Sacha inchi oil	*Plukenetia volubilis*	46
Camelina oil	*Camelina sativa*	38
Rosehip oil	*Rosa Rubiginosa*	28
Canola oil / Rapeseed oil	*Brassica campestris*	9
Soybean Green raw	*Glycine max*	0.4
Soybean oil		6.8
Walnuts	*Juglans regia*	9.1
Cloudberry	*Rubus chamaemorus*	1.2
Blueberry	*Vaccine corymbosum*	0.8
Lingonberry/Cowberry	*Vaccinium vitis-idaea*	0.2

There are few natural sources that have significant amounts of ALA to meet the consumers demands (Table 1) (Valenzuela et al., 2011; Morales et al., 2012). Flaxseed oil is widely used industrially in the manufacture of varnishes and paints for its drying effect but it is used very scarcely as food due to its high oxidative unstability (Valenzuela et al., 2014).

Recently, other sources of ALA are being proposed experimentally, like chia (*Salvia Hispanica*) or sacha inchi (*Pluketia volubilis*) among others. These oils, which have Latimamerican origin, together with others of European and/or Asian origin (camelina, perilla, etc.), are currently interesting alternatives to provide ALA (Valenzuela et al., 2014). In particular, perilla oil (approximately 55% ALA), although its consumption is restricted to Asia, while the consumption of camelina oil (38% ALA) is mainly consumed in Nordic countries. (Barceló-Coblijn et al., 2009; Morales et al., 2012).

3. Bioconversion of ALA to EPA and DHA

In mammals, ALA, after being absorbed at the intestinal level, presents three possible metabolic pathways: storage in adipose tissue, β-oxidation and extensive carbon recycling and conversion to EPA and DHA. β-Oxidation constitutes the main metabolic destination of ALA, approximately between 60-85%, so that there is little ALA available for bioconversion in EPA and DHA. However, this high percentage of ALA used for β-oxidation is not exclusive to ALA, since other PUFA derived from the diet are β-oxidized in similar percentages. For instance, approximately 50% of n-6 fatty acids are used for β-oxidation and about 65% of DHA suffer the same fate. Therefore, dietary ALA, similar to other fatty acids in the diet, undergoes significant utilization by tissues in the form of energy. Finally, the carbon recycling of ALA is quite efficient, being used to form saturated fatty acids (Barceló-Coblijn et al., 2009).

ALA is converted to EPA and DHA through a series of chain desaturation and elongation processes, as stated above. This enzymatic conversion is quite inefficient, mainly concerning to the production of DHA. In particular, it is estimated that the conversion efficiency of ALA to EPA is 0.2%, to DPA n-3 of 0.13% and to DHA of 0.05% (Sanz-París et al., 2012). As shown in Figure 2, in the endoplasmic reticulum, ALA is desaturated at 18:4 n-3 by Δ-6 desaturase, then elongated to 20:4 n-3, and finally converted to EPA by Δ-5-desaturase. EPA can be further elongated to form 22:5 n-3 and then to 24:5 n-3 followed by a desaturation by Δ-6 desaturase to form 24:6 n-3. Subsequently, in the peroxisome, 24:6 n-3 is subjected to a β-oxidation to form DHA (Barceló-Coblijn et al., 2009; Valenzuela et al., 2011).

On the other hand, n-3 and n-6 fatty acids use the same enzymatic machinery for their elongation and desaturation, and thus, there is a competitive relationship between the two routes. The activation of one route or another will depend on the physiological situation of the individual. To better understand this competence, it must be taken into account that the family of n-6 fatty acids has LA as a precursor and one of its most important derivatives is arachidonic acid (20:4 n-6, AA). ALA appears to be a much stronger suppressant of the elongation and desaturation of n-6 fatty acids than the LA of elongation and desaturation of n-3 PUFA.

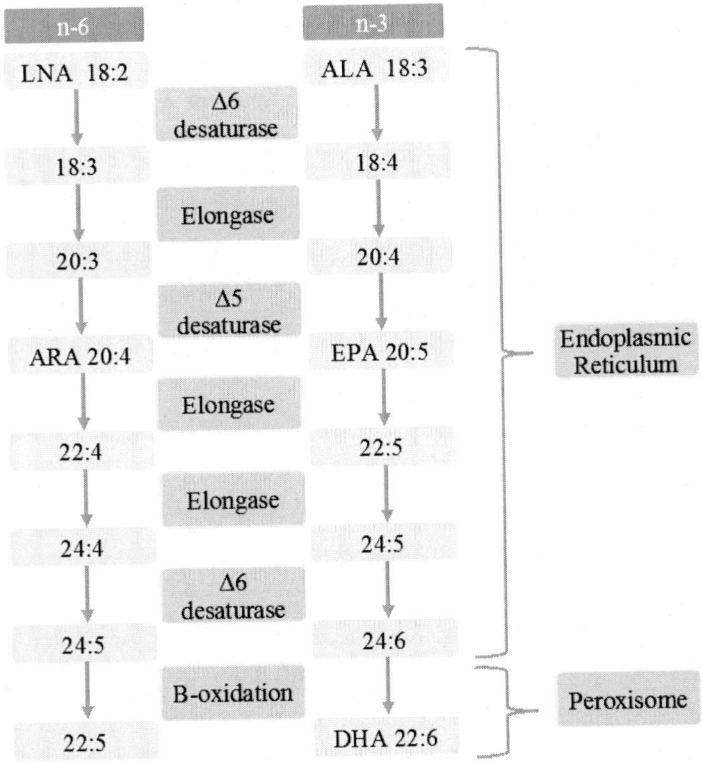

Figure 2. Elongation and desaturation pathways for n-3 and n-6 fatty acid families.

Elongation and desaturation of fatty acids is subject to feedback regulation because both AA and DHA suppress the endogenous conversion of LA and ALA into longer chain fatty acids, respectively (Barceló-Coblijn et al., 2009; Valenzuela et al., 2011). Whereas increasing the level of dietary ALA provides more substrate for conversion to EPA, high levels of dietary ALA also have the capacity to inhibit one of the last steps, the desaturation of 24:5 n-3 (Tu et al., 2010). It should also be considered that the modulation of fatty acids from these reactions appears to be tissue-selective, since dietary n-3 fatty acids are more efficiently elongated and desaturated in the liver, while the brain does not respond to dietary levels and the heart does not have the ability to elongate and desaturate ALA in longer chain n-3 fatty acids. (Igarashi et al., 2007; Barceló-Coblijn et al., 2009)

Several studies carried out in humans have shown that in the metabolism of ALA there is no dose-dependent relationship for the biosynthesis of EPA and DHA from this precursor. (Pawlosky et al., 2001; Schwab et al., 2006; Whelan et al., 2012). But although the conversion of ALA to DHA occurs in a very low proportion, it is still important, and much less negligible, given that an even increased accumulation of DHA, specifically in the brain, is achieved (Morales et al., 2012).

4. Health Effects of ALA

A number of scientific studies have shown the diverse clinical applications of n-3, mainly EPA and DHA, highlighting their benefits in cardiovascular health, dyslipidemias, cancer, diabetes mellitus, rheumatoid arthritis, psychiatric and neurodegenerative diseases or inflammatory bowel disease among others. The results of these trials led international organizations to increase recommendations for the use of EPA and DHA both for the prevention and treatment of these diseases (Aires et al., 2005; Morales et al., 2012; Valenzuela et al., 2013; Zarrouk et al., 2017). In contrast, much less is known about the effects of ALA directly, or after bioconversion to EPA and DHA. Still, some trials have been carried out and are summarized below.

4.1. Preterm and Newborns

ARA and DHA are the most abundant PUFA in the central nervous system, being particularly concentrated in the synaptic plasma membranes and photoreceptor cells. For this reason, it is essential that the correct acquisition of n-3 and n-6 fatty acids occur during embryogenesis and the early postnatal stages of development. The rapid accumulation of DHA is directly related to its crucial need for normal neurological and visual development, in such a way that the fatty acid composition of infant milk is the main object of study. The composition of breast milk varies according to diet, but on average it contains DHA (0.3-0.6%), AA (0.4-0.7%), LNA (8-17%) and ALA (0.5-1%). It should be taken into account that the Adequate Intake in the FAO/WHO recommendations from 2008 was between 0.2 and 0.4% of fatty acids (FAO, 2008). Regarding the ratio between n-6 and n-3 fatty acids, or LNA/ALA, the FAO report concluded that there was no compelling scientific rationale for recommending a specific ratio. However, in the global CODEX standards for infant formula, it is recommended that the LA/ALA ratio be between 5 and 15 (Koletzko et al., 2005). Concerning infant feeding, the evidence provided by human studies indicates that ALA-enriched milk formula have a modest impact on DHA levels. Thus, infants taking these formulas have lower levels of DHA than those breastfed. Therefore, the addition of AA and DHA to infant formulas has been recommended (Clandinin et al., 1981; Koletzko et al., 2005).

4.2. Cardiovascular Disease

Some of the potential mechanisms for the cardioprotective effect of n-3 fatty acids include antiarrhythmic, antiinflammatory, hypotensive, and hypotriglyceridemic effects. However, it is still unclear if all n-3 fatty acids could have beneficial effects on CVD or if the benefit is exclusively related to EPA and DHA. Therefore, there has been great interest in investigating whether sources enriched with ALA would have similar effects to sources enriched with DHA in CVD but the results of these studies are not as consistent as studies with fish oil diets (Geleijnse et al., 2010).

Studies in humans show that within three hours after its consumption, ALA is rapidly incorporated into lipoproteins and, consequently, the plasma concentrations of ALA, EPA and DPA are increased. In a randomized, double-blind trial, 56 participants received 3g ALA/day in the form of flaxseed oil capsules or olive oil placebo capsules. After 12 weeks of treatment, plasma levels of EPA and DPA increased 60% and 25% respectively, but no changes were observed in the concentrations of triglycerides, HDL and LDL nor in the particle size of LDL, HDL or IDL (Layne et al., 1996).

Some authors suggest that the reason for these limited effects may reflect the limited ability of humans to elongate and desaturate ALA. However, a recent study individually evaluated the different metabolic effects of isolated dietary ALA, EPA and DHA in non-obese normolipidemic volunteers. In this study, volunteers received margarines for 6 weeks that contained 4.4g/day of ALA, 2.2g/day of EPA or 2.3g/day of DHA. The results showed not only an increase in the content of ALA and EPA in LDL, but also a reduction in fasting serum triglycerides in the ALA group. Importantly, as noted by the authors, this beneficial effect was observed after the consumption of a relatively low dose of ALA, which could be perfectly achievable without dietary supplements, through regular consumption of ALA-rich foods (Egert et al., 2009). In addition, there are several clinical trials and systematic reviews that attribute beneficial effects to ALA diets because they reduce the risk of myocardial infarction and fatal ischemic heart disease in women and men (Djoussé et al., 2001; Hu et al., 2002).

n-3 PUFA also have antiarrhythmic effects, which probably involve the modification of ion channel currents by the incorporation of these fatty acids into the phospholipid membrane of cardiomyocytes. ALA, EPA and DHA, can block the specific Kv1.5 channels of the atrium at physiological concentrations, suggesting that ALA could share the antiarrhythmic properties of marine n-3 PUFA. (Barceló-Coblijn et al., 2009).

4.3. Inflammatory Diseases

Inflammation is a key feature in a series of clinical conditions, such as cardiovascular diseases, neurodegenerative diseases, cancer, chronic intestinal inflammation, rheumatoid arthritis and asthma among many other pathologies. Inflammation is the cellular physiological response to exposure to certain substances released, predominantly, by activated leukocytes. This response usually consists of the reddening and swelling of the target area and is mediated by at least two different groups of biomolecules, eicosanoids and cytokines. As we saw in the previous section, the increase in EPA and DHA levels, reduce the ARA content in cell membranes and decrease the generation of proinflammatory eicosanoids derived from n-6 PUFA (Valenzuela et al., 2011). One of the most common pharmacological approaches to treat inflammation is to inhibit the biosynthesis of these n-6 eicosanoids, which could be achieved by providing COX1/2 inhibitors or by increasing the content of n-3 fatty acids, particularly EPA and DHA through diet.

Diets enriched with n-3 PUFA can also reduce the production of proinflammatory cytokines, such as interleukin-1, interleukin-6, interleukin-8 and tumor necrosis factor-α (TNF-α), that are released when macrophages and monocytes are activated. In such way, both ALA and DHA can suppress the expression of proinflammatory cytokines by means of a reduction in gene expression induced by the nuclear factor κB (NF-κB) and the peroxisomal proliferation factors (PPARs) (Siebenlist et al., 1994; Zhao et al., 2005). Furthermore, this effect would be related to the cardioprotective effects of ALA, which would be, at least, partially mediated by a reduction in the production of inflammatory cytokines (Zhao et al., 2007).

4.4. Neuroprotective Effect

The protective effects of n-3 PUFA against neurodegenerative disorders, such as Alzheimer's disease are very limited, but have gained attention due to the fact that DHA is a key component of membrane phospholipids in the brain. Moreover, observational studies reported lower incidences of Alzheimer's disease in populations who consume a high amount of fish (Gregg et al., 2002; Morris et al., 2003). A prospective study conducted in participants aged 65–94 y found that ALA intake was strongly protective in persons with a high risk of Alzheimer's disease (Morris et al., 2003).

However, for the moment, most of the available data on the effect of ALA on neurodegenerative disorders is restricted to experimental animals. Yamamoto et al. (1987) reported increased learning ability with supplementation of ALA-rich oil in rats, but most of the effects have been observed after supplementation of the fatty acid individually and not in the context of a diet. Administration of ALA to brain and spinal chord ischemic rats showed significant neuroprotection by reducing neuronal cell loss, inhibiting of microglia activation and improving functional outcome (King et al., 2006; Liu et al., 2014).

4.5. Cancer

In studies carried out with cell cultures and experimental animals, it has been found that diets containing EPA and DHA delay both the growth, the metastasis of primary tumors and the human carcinoma implants in mammary cells. In addition, they induce apoptosis and cell differentiation, as well as the reduction of cell proliferation (Wu et al., 2005). In rats with colon cancer, n-3 PUFA have been shown to block drug-induced tumor formation and directly affect the expression of genes related to tumorigenesis and apoptosis, reducing cellular and DNA damage (Yam et al., 2001).

Doxorubicin, a conventional chemotherapy agent, has limited clinical use due to its systemic toxicity to normal healthy tissue. However, a new conjugate of doxorubicin with ALA showed a good antitumor activity with lower toxicity than free doxorubicin. In addition, the conjugate exhibited a more active tumor targeting profile due to the introduction of ALA which could be an effective tumor directed remnant for the modification of chemotherapeutics (Huan et al., 2012).

However, there is also negative evidence regarding ALA. Christensen et al., (2006) provided support for a deleterious role of ALA in the development of prostate cancer. In 20 men with prostate cancer, these authors did not find any beneficial effects of EPA and DHA intake, but more remarkably, they reported a correlation between the prostate-specific antigen level and ALA concentrations in prostate tissue. Nevertheless, a meta-analysis failed to confirm an association between dietary ALA intake and prostate cancer risk (Carleton et al., 2013). In contrast, there seems to be more evidence on the influence of ALA on breast cancer (Mason and Thompson, 2014). In mice, dietary flaxseed oil was shown to reduce human estrogen receptor-positive breast tumors (MCF-7) growth by 33% compared to control (Truan et al., 2010). In humans, Thompson et al., (2005) demonstrated that a daily intake of 25g flaxseed oil can reduce cell proliferation and increase cell apoptosis in tumors of postmenopausal patients with breast cancer.

The mechanisms of the putative protective effect of ALA, but also EPA and DHA, against cancer can be related to inhibition of tumor development by eliciting changes in the cell membrane fatty acid composition, suppression of ARA-derived eicosanoid biosynthesis, inhibition of cell proliferation and induction of apoptosis (Liu and Ma, 2014).

5. Other Alternative Sources Of n-3 PUFA

The recommended daily intakes of n-3 PUFA are often viewed as unsustainable and unable to meet future consumer demands (Tocher, 2009). The global demand of wild fish, the most common and effective source of EPA and DHA, is much larger than the oceans can sustain (Arthur, 2009). The concern about this demand has moved efforts to develop alternative sources of n-3 PUFA, including aquaculture, other marine non-fish products, microalgae and transgenic plants.

Aquaculture is an increasing food industry, with an annual average of 8.3% increase (Klinger and Naylor, 2012). However, in order to reach high n-3 PUFA levels in their flesh, seafood and fish grown in aquaculture require formulated feeds composed of wild caught fish or EPA and DHA supplementation when being farmed (Adarme-Vega et al., 2014). ALA-containing plants, like soybean, canola and flaxseed, have also been suggested as additives for fish food. However, although fish fed plants present mostly positive effects on growth performance, their meat results in lower accumulation of n-3 PUFA, so consumers do not gain the same health benefits. Therefore, aquaculture alone is not sufficient to solve the problem of n-3 PUFA demand.

Since ALA-rich plants, which as we have reviewed above, are an unrealistic alternative to meet EPA and DHA demands, krill oil, microalgae and genetically modified plants have also been suggested as sources of n-3 PUFA.

5.1. Krill

Krill are small crustaceans particularly abundant in the Arctic and Antarctic oceans. The oil is extracted from Euphausia superba, the largest krill species that lives in the Antartic. However, due to their role in marine ecosystems, Antartic krill harvest is limited to assure for long-term sustainability (Constable, 2011). Therefore, considerable caution is needed when increasing the krill catches as it compromises the marine food web.

Krill lipid content is variable (12-50% of dry weight) as it depends on the season, species and age. However, it is particularly rich in EPA and DHA, which is attributed to their diet based on microalgae. The overall fatty acid composition in krill oil is similar to fish oil but the EPA content is higher, increasing the EPA:DHA ratio from 1:1 in fish oil to 2:1 in krill oil (Ulven et al., 2011). This might have an important impact on eicosanoid synthesis and their biological response as stated above.

In contrast to fish, in which EPA and DHA are found in triglycerides, in krill these fatty acids are mainly found as part of phospholipids (mostly in the form of phophatidylcholine) (Ulven et al., 2011). This is of particular interest, since in has been demonstrated that there is about two-fold higher DHA accumulation in the brain when this fatty acid is provided as phospholipids rather than triglycerides (Liu et al., 2014).

Phospholipids have an amphiphilic nature and it has been suggested that their emulsifying features might contribute to a better absorption in the intestine (Schuchardt and Hahn, 2013). However, this statement is controversial and has been questioned by some authors. Köhler et al., (2015) carried out a randomized, single-blind, cross-over intervention trial to compare the bioavailability of EPA and DHA administrated from Antarctic krill oil, Antarctic krill meal and fish oil. These authors found a higher bioavailability of EPA and DHA from krill oil but no differences between krill meal and fish oil, which argues against the interpretation that phospholipids are better absorbed than triglycerides. Yurko-Mauro et al., (2015) performed a double blind, randomized, parallel study in which 66 healthy subjects received a 1.3 g/day dose of EPA + DHA in the form of fish oil-ethyl esters, fish oil-triglycerides or krill oil for four weeks. They were unable to find differences in plasma and erythrocyte fatty acid composition, demonstrating that all forms of EPA + DHA had similar oral bioavailability.

However, studies on the bioavailability of EPA and DHA from krill oil are conflicting due to differences in design, including the amounts of EPA and DHA provided by krill or fish oil. Studies that used the same amount of EPA and DHA showed that the bioavailability of EPA and DHA from krill oil seems to be higher compared to fish oil (Ramprasath et al., 2013; Schuchardt et al., 2011). On the contrary, others found no difference (Laidlaw et al., 2014). Despite using the same amounts of EPA and DHA, these studies are difficult to compare because of divergences in duration but also because fatty acids were analyzed in different plasma matrices: whole plasma, plasma phospholipids and erythrocytes. Nevertheless, studies in humans that assessed the biological effect of krill oil and fish oil did not find significant differences (Ulven et al., 2015). In contrast, studies in animals show that fish oil intake, unlike krill oil, upregulates the cholesterol synthesis pathway. In addition, krill oil seems to regulate also other glucose and lipid metabolic pathways.

Fish oil and krill oil have also been compared in regard to cognitive function in senior citizens (61-72 y) (Konagai et al., 2013). Subjects were asked to take four capsules of each oil and a placebo (medium-chain triglycerides) twice a day for 12 weeks. Again, no differences were found between fish and krill oils. Both oils increased oxyhemoglobin concentrations in the brain, which is associated with enhanced cerebral function related to circulation. In consequence, the authors concluded that long-term ingestion of krill or fish oil promotes working memory function by activating the dorsolateral prefrontal cortex in elderly people, preventing deterioration in cognitive activity. However, only krill oil had improving effects in calculation tasks performed by the participants, which was associated to the presence of EPA and DHA in phospholipids.

Another randomized crossover trial used a combination of krill oil (88%) and salmon oil (12%), containing contained 46 mg EPA and 31 mg DHA (Albert et al., 2015). Participants were instructed to take 400 mg n–3 PUFA as capsules per day for eight weeks. Remarkably, the authors found a decrease in insulin sensitivity compared to the control (canola oil), increased LDL-cholesterol and apolipoprotein B concentrations and carotid artery intima-media thickness. This suggested a more atherogenic lipid profile and higher risk of type-2 diabetes due to krill oil consumption. The authors were reluctant to attribute the effects to EPA and DHA or to astaxanthin, an antioxidant carotenoid, contained in the manufactured capsules, but were unable to provide another explanation.

Krill oil has also been shown to reduce plasma triglycerides large very-low density lipoprotein (VLDL) and chylomicron particles in healthy individuals after consumption for 28 days, which was associated to to increased plasma EPA, DHA and docosapentaenoic acid (DPA) concentrations and reduced ARA levels (Berge et al., 2014). Plasma EPA and DHA concentrations were also shown to be increased in overweight and obese men and women receiving capsules containing 2 g/day of krill oil for 4 weeks (Maki et al., 2009). In addition, patients treated for three months with 1 g/day of krill oil showed reduced total cholesterol, LDL, and triglycerides, and increased HDL levels, with no difference with fish oil (Bunea et al., 2004). These studies demonstrated that krill oil can be at least as effective as fish oil for the management of hyperlipidemia.

5.2. Algae

Algae have been part of the human diet for thousands of years, but more contemporaneously, in addition to macroalgae, some microalgae are cultivated for foods and food additives (Chacón-Lee and González-Mariño 2010; FAO, 2016). Today, algae are regarded as important sources of n-3 PUFA, particularly for vegetarians, who may have low intakes of these fatty acids. Microalgae are the primary sources of DHA and EPA for zooplankton and fish. Thus, microalgae represent the trophic integration of DHA-rich flagellates and EPA-rich diatoms in the food web (Viso and Marty, 1993). Although algae have been proposed as alternatives sources to fish oil for n-3 PUFA intake because they are easy to cultivate on a large scale due to their small size, ocean acidification, as a result of increased atmospheric CO_2, can negatively affect the supply of these fatty acids to higher trophic levels (Rossoll et al., 2012).

Among the ca. 40,000 algal species that have been identified, only a few hundred have been characterized for their lipid composition (Adarme-Vega et al., 2012). Despite algae exhibit a great variability regarding their lipid content, some microalgae can accumulate up to 60% of their dry weight as triglycerides (Georgianna and Mayfield, 2012). EPA is the predominant PUFA in many of them, along with DHA and ARA (Norziah and Ching, 2000). Some strains have demonstrated high accumulation of EPA and/or DHA. For instance, species of *Phaeodactylum* and *Nannochloropsis* can reach up to 39% of EPA (Yongmanitchai and Ward, 1991; Sukenik, 1991), whereas *Thraustochytrium sp.* and *Schizochytrium sp.* (Burja et al., 2006; Zhu et al., 2007) contain DHA in amounts ranging 30–40% of total fatty acids when grown heterotrophically. The bioaccessibility for these fatty acids ranges from 50 to 100% (Schuchardt et al., 2011).

EPA and DHA content can be modified by changing growing conditions, such as temperature. For instance, phytoplankton contain more PUFA when grown at low temperature (Yue and Feng, 2000). In addition, low nitrogen conditions with high supply of monosodium glutamate can improve growth and fatty acid synthesis (Lenihan-Geels et al., 2013). Under stressful low nutrient conditions (but not light), microalgae tend to accumulate and store photosynthetic products, such as lipids and carbohydrates (Anderson et al., 2008). Apart for external stresses, metabolic engineering has been proposed to increase the lipid production of microalgae, by regulating the expression of desaturases and elongases that are needed for n-3 PUFA biosynthesis (Adarme-Vega et al., 2012). Additionally, other options include the modulation of the enzymes involved in degradation pathways, like β-oxidation.

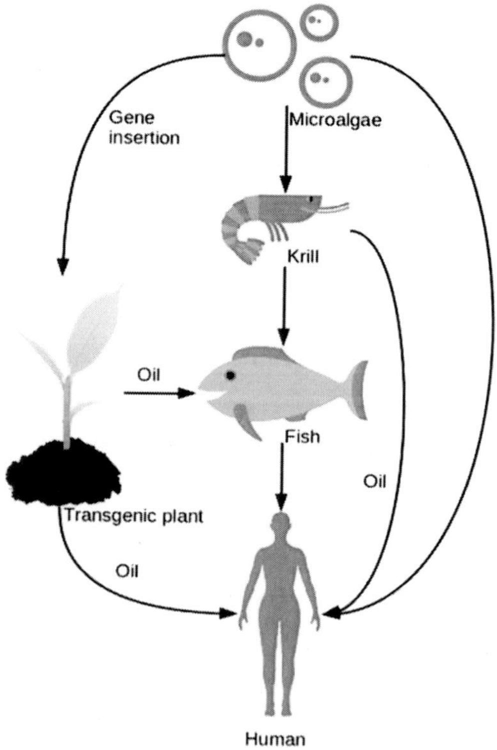

Figure 3. Microalgae, krill and transgenic plants as n-3 PUFA sources in human trophic chain.

As indicated above, algal oil can be particularly interesting for vegetarians because these individuals are at risk of a lower status of n-3 PUFA, due to the exclusion of fish from their diet (Lee et al., 2000). In addition, they usually have a high intake of n-6 PUFA from vegetable oils, mainly LA, which compete with the conversion of ALA to EPA and DHA. In a randomized controlled trial, postmenopausal vegan and lacto-vegetarian women received 2.14 g of DHA/day for 6 weeks and no EPA from algal oil. The study showed that microalgal oil increased the content of DHA, but also EPA in LDL, in the form of cholesterol esters. Compared to the control, plasma total cholesterol concentrations were slightly reduced but no significant changes were observed in LDL and HDL. Likewise, plasma triglyceride levels were reduced but not significantly. The authors suggested that this poor effect was related to the low daily dose of n-3 fatty acids and original favorable lipid profiles in vegetarians.

However, algal DHA can have cardiovascular protective effects by exerting changes in lipoprotein concentrations when administered at small doses (Neff et al., 2011). A double-blind, randomized, parallel trial included algal oil from *Schizochytrium sp.*, fish oil or a corn/soy oil as control (Maki et al., 2014). The oils were administered in the form of capsules, providing 443 mg DHA and 164 mg EPA per capsule in the case of algal oil, and 205 mg DHA and 289 mg EPA for fish oil. Healthy adult men and women were treated with four capsules per day for 14 weeks. Ingestion of microalgal oil significantly lowered triglyceride levels compared with the control, but it was not significantly different from that produced by fish oil. This lowering effect of microalgal and fish oils is concomitant to reduced concentrations of VLDL triglycerides as well as VLDL particle size (Neff et al., 2011). Interestingly, the trial by Maki et al. (2014) also found that both microalgal and fish oils significantly increased LDL compared with control. However, this effect was associated with a shift of lipoprotein particle size toward larger, potentially less atherogenic subfractions. Nevertheless, the health benefits of DHA-rich algal supplements are controversial when compared to fish, probably due to heterogeneity in design of the studies in terms of amounts administered, duration and sample size, just as we showed above for krill oil.

5.3. Transgenic Plants

Since terrestrial plants do not produce very long-chain n-3 PUFA because they lack the necessary genes required to elongate and desaturate ALA, the only option to obtain EPA and DHA from plants is genetic modification. Transgenic plants have been proposed as novel terrestrial sources of n-3 PUFA. It has been argued that these plants represent a less expensive and more sustainable alternative to microalgae culture and fish farming, and can avoid well-known concerning issues associated with ocean fishing, such as contamination with heavy metals, dioxins and polychlorinated biphenyls (Tocher, 2015). Oils derived from these transgenic plants could be directed to human consumption, but also as aquafeed ingredient, as a replacement for fish oils.

The first attempt to generate a plant rich in EPA and DHA was made in *Arabidopsis thaliana,* via insertion of microbial desaturases and elongases enzymes, and achieving the formation of 3% EPA in the leaf tissues (Barclay et al., 1994). More recently, Petrie et al. (2012) described the accumulation of DHA up to 15%, also in genetically modified *Arabidopsis thaliana,* but the EPA content was even lower (1.8%). Today, new sources of n-3 PUFA have been developed from *Camelina sativa*, a commercially harvested plant that can synthesize EPA and/or DHA when it is genetically modified (Ruiz-Lopez et al., 2014). Transgenic Camelina is capable of producing n-3 PUFA in their seeds after the insertion of cassettes containing a number of fatty acyl desaturase and elongase genes from algae. Following this procedure, seeds can contain up to 31% EPA or 12% EPA and 14% DHA.

These oils have been evaluated as replacements for fish oil in feeds for salmon (*Salmo salar*) (Betancor et al., 2015) and gilthead sea bream (*Sparus aurata*) (Betancor et al., 2016). In these trials growth performance and n-3 PUFA deposition in flesh was comparable to fish fed diets containing wild-type camelina oil and no adverse effects were observed on plasma biochemistry or intestinal histology. Although levels of n-3 PUFA did not reach the concentrations in flesh that were achieved when fish received diets containing fish oil as lipid source, camelina-derived oils doubled the n-3 PUFA content in flesh, in comparison with commercial feeds (Betancor et al., 2017). This effect was attributed to up-regulation of hepatic fads2d5 and fads2d6 genes, corresponding to delta-5 and delta-6 fatty acyl desaturases, respectively.

When EPA-rich or EPA+DHA-rich camelina oils were compared, Betancor et al. (2016) observed reduced performance in fish growth after the administration of the former that was associated to an imbalance between EPA and DHA content. In addition, these authors found increased ARA content in the meals, which is associated to pro-inflammatory effects, as indicated above. Nevertheless, they observed no significant differences in flesh fatty acid composition, and no alterations in fish health as shown by histological and genetic expression analyses.

Fish fed the transgenic oil presented increased lipid deposition in the liver, which had already been observed in fish fed plant oils, compared to those fed marine fish oil (Fountoulaki et al., 2009). An explanation for this observation is that n-3 PUFA have been shown to suppress triglyceride accumulation in mammalian and fish adipocytes (Kim et al., 2006; Todorcević et al., 2008). An alternative explanation is related to enhanced expression in srbep-1, which in turn regulates the expression of hepatic and lipoprotein lipases. These enzymes are in charge of triglyceride hydrolysis from lipoproteins and, thus, are involved in the hepatic uptake of fatty acids.

For the moment, there are no trials in humans consuming dietary fish farmed using genetically modified plant oils, due to the strict regulations of transgenic products and low consumer acceptance in the European Union. Transgenic plants have an enormous potential as source of n-3 PUFA, but, in the present circumstances, they can not be considered in the short term.

CONCLUSION

n-3-PUFA are recognized as highly beneficial for human health through their effects on the most prevalent non-communicable chronic diseases: cardiovascular disease, diabetes, cancer and neurodegenerative disorders. International health organizations recommend increasing the intake of fish and fish oils, the most common and richest sources of EPA and DHA, which has led to a growing interest in the general population. Therefore, the market demand for n-3 PUFA has increased in the last years and it is expected that it will further increase in the future. This not only implies direct intake by humans in the form of fish or fish oil, but also the production of nutraceuticals and animal feed. Consequently, it has become urgent to find new sustainable sources of n-3 PUFA to meet that demand. This is particularly imperative in certain groups of populations, like vegetarians, who do not eat n-3 PUFA from animal sources. In addition, marine fish stocks have declined in the last decades and overfished stocks have increased. Moreover, the presence of chemical contaminants (e.g., mercury) in fish oil are of great concern for most consumers.

Aquaculture was proposed as a solution for the problem but it implies that farmed fish need to be fed with wild caught fish. Therefore, aquaculture alone is not sufficient to solve the problem of n-3 PUFA demand. Plant oils, like flaxseed, chia or sacha inchi oils, but also some nuts, are rich sources of ALA and have proven to be highly bioavailable as alternative sources to fish. However, bioconversion of ALA to EPA and DHA is rather poor, leading scientists to search for other sources of n-3 PUFA, in the form of algae, krill or transgenic plants. Cultivated microalgae are grown in cost-effective processes and it is likely to be an expanding industry in the future. However, the industrialized processes need to be enhanced if these source of n-3 PUFA is to be dedicated to replace a significant part of fish intake. Krill oil is another proposed source of n-3 PUFA but it is unlikely that it can be sufficiently sustainable. Finally, transgenic plants represent a potentially more sustainable source which also avoid issues associated with marine sources, such as contamination with heavy metals, dioxins and PCBs. In this case, concerns are related to social pressure, since genetically modified organisms are not well accepted by consumers and policy makers. It is likely that it will be not one, but a combination of the above sources which provide a future sustainable production of n-3 PUFA.

REFERENCES

Adarme-Vega, T. C., Thomas-Hall, S. R. and Schenk, P. M. (2014). Towards sustainable sources for omega-3 fatty acids production. *Current Opinion in Biotechnology*, 26: 14–18. https://doi.org/10.1016/j.copbio.2013.08.003.

Adarme-Vega, T., Lim, D. K. Y., Timmins, M., Vernen, F., Li, Y. and Schenk, P. M. (2012). Microalgal biofactories: a promising approach towards sustainable omega-3 fatty acid production. *Microbial Cell Factories*, 11(1): 96. https://doi.org/10.1186/1475-2859-11-96.

Aires, D., Capdevila, N. and Segundo, M. J. (2005). Ácidos Grasos Esenciales. [Essential fatty acids.] *www.Nutri-Facts.Org*, 24: 96–102. https://doi.org/10.1590/S1415-790X2003000400003.

Albert, B. B., Derraik, J. G., Brennan, C. M., Biggs, J. B., Garg, M. L., Cameron-Smith, D. et al., (2015). Supplementation with a blend of krill and salmon oil is associated with increased metabolic risk in overweight men. *American Journal of Clinical Nutrition*, 102(1): 49–57. https://doi.org/10.3945/ajcn.114.103028.

Anderson, D. M., Burkholder, J. M., Cochlan, W. P., Glibert, P. M., Gobler, C. J., Heil, C. A. et al., (2008). Harmful algal blooms and eutrophication: Examining linkages from selected coastal regions of the United States. *Harmful Algae*, 8(1): 39–53. https://doi.org/10.1016/j.hal.2008.08.017.

Arthur, R. (2009). Omega-3 sources. *Journal of Complementary Medicine*, 8: 28.

Azrad, M., Turgeon, C., Demark-Wahnefried, W. (2013). Current evidence linking polyunsaturated fatty acids with cancer risk and progression. *Front Oncol,* 3:224.

Barber, M. D., Ross, J. A., Voss, A. C., Tisdale, M. J., Fearon, K. C. (1999). The effect of an oral nutritional supplement enriched with fish oil on weight loss in patients with pancreatic cancer. *Br J Cancer*. 81:80–6.

Barceló-Coblijn, G., Kitajka, K., Puskás, L. G., Hőgyes, E., Zvara, A., Hackler, L. and Farkas, T. (2003). Gene expression and molecular composition of phospholipids in rat brain in relation to dietary n−6 to n−3 fatty acid ratio. *Biochimica et Biophysica Acta (BBA) - Molecular and Cell Biology of Lipids*, 1632(1–3): 72–79. https://doi.org/10.1016/S1388-1981(03)00064-7.

Barceló-Coblijn, G. and Murphy, E. J. (2009). Alpha-linolenic acid and its conversion to longer chain n−3 fatty acids: Benefits for human health and a role in maintaining tissue n−3 fatty acid levels. *Progress in Lipid Research*, 48(6): 355–374. https://doi.org/10.1016/J.PLIPRES.2009.07.002.

Barclay, W. R., Meager, K. M. and Abril, J. R. (1994). Heterotrophic production of long chain omega-3 fatty acids utilizing algae and algae-like microorganisms. *Journal of Applied Phycology*, 6(2): 123–129. https://doi.org/10.1007/BF02186066.

Berge, K., Musa-Veloso, K., Harwood, M., Hoem, N. and Burri, L. (2014). Krill oil supplementation lowers serum triglycerides without increasing low-density lipoprotein cholesterol in adults with borderline high or high triglyceride levels. *Nutrition Research*, 34(2): 126–133. https://doi.org/10.1016/j.nutres.2013.12.003.

Berquin, I. M., Edwards, I.J., Chen, Y. Q. (2008). Multi-targeted therapy of cancer by omega-3 fatty acids. *Cancer Lett,* 269:363–77.

Betancor, M. B., Li, K., Sprague, M., Bardal, T., Sayanova, O., Usher, S. et al., (2017). An oil containing EPA and DHA from transgenic Camelina sativa to replace marine fish oil in feeds for Atlantic salmon (Salmo salar L.): Effects on intestinal transcriptome, histology, tissue fatty acid profiles and plasma biochemistry. *PLOS ONE*, 12(4): e0175415. https://doi.org/10.1371/journal.pone.0175415.

Betancor, M. B., Sprague, M., Montero, D., Usher, S., Sayanova, O., Campbell, P. J. et al., (2016). Replacement of Marine Fish Oil with de novo Omega-3 Oils from Transgenic Camelina sativa in Feeds for Gilthead Sea Bream (Sparus aurata L.). *Lipids*, 51(10): 1171–1191. https://doi.org/10.1007/s11745-016-4191-4.

Betancor, M. B., Sprague, M., Usher, S., Sayanova, O., Campbell, P. J., Napier, J. A. and Tocher, D. R. (2015). A nutritionally-enhanced oil from transgenic Camelina sativa effectively replaces fish oil as a source of eicosapentaenoic acid for fish. *Scientific Reports*, 5(1): 8104. https://doi.org/10.1038/srep08104.

Bougnoux, P., Hajjaji, N., Ferrasson, M. N., Giraudeau, B., Couet, C., et al., (2009). Improving outcome of chemotherapy of metastatic breast cancer by docosahexaenoic acid: a phase II trial. *Br J Cancer,* 101:1978–85.

Bunea, R., El Farrah, K. and Deutsch, L. (2004). Evaluation of the effects of Neptune Krill Oil on the clinical course of hyperlipidemia. *Alternative Medicine Review : A Journal of Clinical Therapeutic*, 9(4): 420–428.

Burja, A. M., Radianingtyas, H., Windust, A. and Barrow, C. J. (2006). Isolation and characterization of polyunsaturated fatty acid producing Thraustochytrium species: screening of strains and optimization of omega-3 production. *Applied Microbiology and Biotechnology*, 72(6): 1161–1169. https://doi.org/10.1007/s00253-006-0419-1.

Carleton, A. J., Sievenpiper, J. L., de Souza, R., McKeown-Eyssen, G. and Jenkins, D. J. A. (2013). Case–control and prospective studies of dietary α-linolenic acid intake and prostate cancer risk: a meta-analysis. *BMJ Open*, 3(5): e002280. https://doi.org/10.1136/bmjopen-2012-002280.

Chacón-Lee, T. L. and González-Mariño, G. E. (2010). Microalgae for "Healthy" Foods-Possibilities and Challenges. *Comprehensive Reviews in Food Science and Food Safety*, 9(6): 655–675. https://doi.org/10.1111/j.1541-4337.2010.00132.x.

Christensen, J. H., Fabrin, K., Borup, K., Barber, N. and Poulsen, J. (2006). Prostate tissue and leukocyte levels of n-3 polyunsaturated fatty acids in men with benign prostate hyperplasia or prostate cancer. *BJU International*, 97(2): 270–273. https://doi.org/10.1111/j.1464-410X.2006.05951.x.

Clandinin, M. T., Chappell, J. E. and Heim, T. (1981). Do low weight infants require nutrition with chain elongation-desaturation products of essential fatty acids? *Progress in Lipid Research*, 20: 901–904.

Colomer, R., Moreno-Nogueira, J. M., Garcia-Luna, P. P., Garcia-Peris, P., Garcia-de-Lorenzo, A., et al., (2007). N-3 fatty acids, cancer and cachexia: a systematic review of the literature. *Br J Nutr*, 97:823–31.

Connor, K. M., SanGiovanni, J. P., Lofqvist, C., Aderman, C. M., Chen, J., et al., (2007). Increased dietary intake of omega-3-polyunsaturated fatty acids reduces pathological retinal angiogenesis. *Nat Med.*13:868–73.

Constable, A. J. (2011). Lessons from CCAMLR on the implementation of the ecosystem approach to managing fisheries. *Fish and Fisheries*, 12: 138–151.

Djoussé, L., Pankow, J. S., Eckfeldt, J. H., Folsom, A. R., Hopkins, P. N., Province, M. A. et al., (2001). Relation between dietary linolenic acid and coronary artery disease in the National Heart, Lung, and Blood Institute Family Heart Study. *The American Journal of Clinical Nutrition*, 74(5): 612–619.

Dyerberg, J., Bang, H. O., Hjorne, N. (1975). Fatty acid composition of the plasma lipids in Greenland Eskimos. *Am J Clin Nutr*, 28: 958–66.

Edwards, I. J., O'Flaherty, J. T. (2008). Omega-3 fatty acids and PPARgamma in cancer. *PPAR Res.*, 2008:358052.

Egert, S., Kannenberg, F., Somoza, V., Erbersdobler, H. F. and Wahrburg, U. (2009). Dietary α-Linolenic Acid, EPA, and DHA Have Differential Effects on LDL Fatty Acid Composition but Similar Effects on Serum Lipid Profiles in Normolipidemic Humans. *The Journal of Nutrition*, 139(5): 861–868. https://doi.org/10.3945/jn.108.103861.

FAO. (n.d.). *The State of World Fisheries and Aquaculture* 2016.

FAO. (n.d.). Interim Summary of Conclusions and Dietary Recommendations on Total Fat; *Fatty Acids From the Joint FAO/WHO Expert Consultation on Fats and Fatty Acids in Human Nutrition*, 10-14 November, 2008, WHO, Geneva Introduction and Definitions.

Fearon, K. C., Barber, M. D., Moses, A. G., Ahmedzai, S. H., Taylor, G. S., et al., (2006). Double-blind, placebo-controlled, randomized study of eicosapentaenoic acid diester in patients with cancer cachexia. *J Clin Oncol,* 24:3401–7.

Fearon, K. C., VonMeyenfeldt, M. F., Moses, A. G., Van Geenen, R., Roy, A., et al., (2003). Effect of a protein and energy dense N-3 fatty acid enriched oral supplement on loss of weight and lean tissue in cancer cachexia: a randomised double blind trial. *Gut.,* 52: 1479–86.

Fountoulaki, E., Vasilaki, A., Hurtado, R., Grigorakis, K., Karacostas, I., Nengas, I. et al., (2009). Fish oil substitution by vegetable oils in commercial diets for gilthead sea bream (Sparus aurata L.); effects on growth performance, flesh quality and fillet fatty acid profile: Recovery of fatty acid profiles by a fish oil finishing diet under fluctuating water temperatures. *Aquaculture*, 289(3–4): 317–326. https://doi.org/10.1016/J.AQUACULTURE.2009.01.023.

Geleijnse, J. M., de Goede, J. and Brouwer, I. A. (2010). Alpha-Linolenic Acid: Is It Essential to Cardiovascular Health? *Current Atherosclerosis Reports*, 12(6): 359–367. https://doi.org/10.1007/s11883-010-0137-0.

Georgianna, D. R. and Mayfield, S. P. (2012). Exploiting diversity and synthetic biology for the production of algal biofuels. *Nature*, 488(7411): 329–335. https://doi.org/10.1038/nature11479.

Gerber, M. (2012). Omega-3 fatty acids and cancers: a systematic update review of epidemiological studies. *Br J Nutr,* 107:S228–39.

Gleissman, H., Johnsen, J. I., Kogner, P. (2010). Omega-3 fatty acids in cancer, the protectors of good and the killers of evil? *Exp Cell Res,* 316:1365–73.

Greene, E. R., Huang, S., Serhan, C. N., Panigrahy, D. (2011). Regulation of inflammation in cancer by eicosanoids. *Prostaglandins Other Lipid Mediat*, 96:27–36.

Gregg, E. W., Engelgau, M. M. and Narayan, V. (2002). Complications of diabetes in elderly people. *BMJ Clinical Research Ed.*, 325(7370): 916–917.

Gu, Z., Suburu, J., Chen, H., Chen, Y. Q. (2013). Mechanisms of omega-3 polyunsaturated fatty acids in prostate cancer prevention. *Biomed Res Int*, 2013:824563.

Gullett, N. P., Mazurak, V. C., Hebbar, G., Ziegler, T. R. (2011). Nutritional interventions for cancer-induced cachexia. *Curr Probl Cancer.* 35:58–90.

Hu, F. B., Bronner, L., Willett, W. C., Stampfer, M. J., Rexrode, K. M., Albert, C. M. et al., (2002). Fish and omega-3 fatty acid intake and risk of coronary heart disease in women. *JAMA*, 287(14): 1815–1821.

Hu, Y., Sun, H., Owens, R. T., Gu, Z., Wu, J., et al., (2010). Syndecan-1-dependent suppression of PDK1/Akt/bad signaling by docosahexaenoic acid induces apoptosis in prostate cancer. *Neoplasia*, 12:826–36.

Huan, M., Cui, H., Teng, Z., Zhang, B., Wang, J., Liu, X. et al., (2012). In Vivo Anti-Tumor Activity of a New Doxorubicin Conjugate *via* α-Linolenic Acid. *Bioscience, Biotechnology, and Biochemistry*, 76(8): 1577–1579. https://doi.org/10.1271/bbb.120256.

Huober, J., Thurlimann, B. (2009). The role of combination chemotherapy in the treatment of patients with metastatic breast cancer. *Breast Care (Basel),* 4:367–72.

Igarashi, M., Ma, K., Chang, L., Bell, J. M. and Rapoport, S. I. (2007). Dietary n-3 PUFA deprivation for 15 weeks upregulates elongase and desaturase expression in rat liver but not brain. *Journal of Lipid Research*, 48(11): 2463–2470. https://doi.org/10.1194/jlr.M700315-JLR200.

Jatoi, A., Rowland, K., Loprinzi, C. L., Sloan, J. A., Dakhil, S. R., et al., (2004). An eicosapentaenoic acid supplement versus Megestrol acetate versus both for patients with cancer-associated wasting: a North Central Cancer Treatment Group and National Cancer Institute of Canada collaborative effort. *J Clin Oncol.,* 22:2469–76.

Kim, H.-K., Della-Fera, M., Lin, J. and Baile, C. A. (2006). Docosahexaenoic acid inhibits adipocyte differentiation and induces apoptosis in 3T3-L1 preadipocytes. *The Journal of Nutrition*, 136(12): 2965–2969.

King, V. R., Huang, W. L., Dyall, S. C., Curran, O. E., Priestley, J. V and Michael-Titus, A. T. (2006). Omega-3 Fatty Acids Improve Recovery, whereas Omega-6 Fatty Acids Worsen Outcome, after Spinal Cord Injury in the Adult Rat. *Journal of Neuroscience*, 26(17): 4672–4680. https://doi.org/10.1523/JNEUROSCI.5539-05.2006.

Klinger, D. and Naylor, R. (2012). Searching for Solutions in Aquaculture: Charting a Sustainable Course. *Annual Review of Environment and Resources*, 37(1): 247–276. https://doi.org/10.1146/annurev-environ-021111-161531.

Köhler, A., Sarkkinen, E., Tapola, N., Niskanen, T. and Bruheim, I. (2015). Bioavailability of fatty acids from krill oil, krill meal and fish oil in healthy subjects–a randomized, single-dose, cross-over trial. *Lipids in Health and Disease*, 14(1): 19. https://doi.org/10.1186/s12944-015-0015-4.

Koletzko, B., Baker, S., Cleghorn, G., Neto, U. F., Gopalan, S., Hernell, O. et al., (2005). Global standard for the composition of infant formula: recommendations of an ESPGHAN coordinated international expert group. *Journal of Pediatric Gastroenterology and Nutrition*, 41(5): 584–599.

Konagai, C., Yanagimoto, K., Hayamizu, K., Li, L., Tsuji, T. and Koga, Y. (2013). Effects of krill oil containing n-3 polyunsaturated fatty acids in phospholipid form on human brain function: a randomized controlled trial in healthy elderly volunteers. *Clinical Interventions in Aging*, 8: 1247. https://doi.org/10.2147/CIA.S50349.

Kroetz, D. L., Zeldin, D. C. (2002). Cytochrome P450 pathways of arachidonic acid metabolism. *Curr Opin Lipidol.,* 13:273–83.

Laidlaw, M., Cockerline, C. A. and Rowe, W. J. (2014). A randomized clinical trial to determine the efficacy of manufacturers' recommended doses of omega-3 fatty acids from different sources in facilitating cardiovascular disease risk reduction. *Lipids in Health and Disease*, 13(1): 99. https://doi.org/10.1186/1476-511X-13-99.

Layne, K. S., Goh, Y. K., Jumpsen, J. A., Ryan, E. A., Chow, P. and Clandinin, M. T. (1996). Normal subjects consuming physiological levels of 18:3(n-3) and 20:5(n-3) from flaxseed or fish oils have characteristic differences in plasma lipid and lipoprotein fatty acid levels. *The Journal of Nutrition*, 126(9): 2130–2140.

Lee, H. Y., Woo, J., Chen, Z. Y., Leung, S. F. and Peng, X. H. (2000). Serum fatty acid, lipid profile and dietary intake of Hong Kong

Chinese omnivores and vegetarians. *European Journal of Clinical Nutrition*, 54(10): 768–773.

Lenihan-Geels, G., Bishop, K. and Ferguson, L. (2013). Alternative Sources of Omega-3 Fats: Can We Find a Sustainable Substitute for Fish? *Nutrients*, 5(4): 1301–1315. https://doi.org/10.3390/nu5041301.

Lewis, R. A., Austen, K. F., Soberman, R. J. (1990). Leukotrienes and other products of the 5-lipoxygenase pathway. Biochemistry and relation to pathobiology in human diseases. *N Engl J Med.*, 323: 645–55.

Liu, J. and Ma, D. W. L. (2014). The role of n-3 polyunsaturated fatty acids in the prevention and treatment of breast cancer. *Nutrients*, 6(11): 5184–5223. https://doi.org/10.3390/nu6115184.

Liu, L., Bartke, N., Van Daele, H., Lawrence, P., Qin, X., Park, H. G. et al., (2014). Higher efficacy of dietary DHA provided as a phospholipid than as a triglyceride for brain DHA accretion in neonatal piglets. *Journal of Lipid Research*, 55(3): 531–539. https://doi.org/10.1194/jlr. M045930.

Liu, Y., Sun, Q., Chen, X., Jing, L., Wang, W., Yu, Z. et al., (2014). Linolenic Acid Provides Multi-cellular Protective Effects After Photothrombotic Cerebral Ischemia in Rats. *Neurochemical Research*, 39(9): 1797–1808. https://doi.org/10.1007/s11064-014-1390-3.

Maki, K. C., Reeves, M. S., Farmer, M., Griinari, M., Berge, K., Vik, H. et al., (2009). Krill oil supplementation increases plasma concentrations of eicosapentaenoic and docosahexaenoic acids in overweight and obese men and women. *Nutrition Research*, 29(9): 609–615. https://doi.org/10.1016/j.nutres.2009.09.004.

Maki, K. C., Yurko-Mauro, K., Dicklin, M. R., Schild, A. L. and Geohas, J. G. (2014). A new, microalgal DHA- and EPA-containing oil lowers triacylglycerols in adults with mild-to-moderate hypertriglyceridemia. *Prostaglandins, Leukotrienes and Essential Fatty Acids (PLEFA)*, 91(4): 141–148. https://doi.org/10.1016/j.plefa.2014.07.012.

Mantovani, G. (2010). Randomised phase III clinical trial of 5 different arms of treatment on 332 patients with cancer cachexia. *Eur Rev Med Pharmacol Sci*, 14:292–301.

Mason, J. K. and Thompson, L. U. (2014). Flaxseed and its lignan and oil components: can they play a role in reducing the risk of and improving the treatment of breast cancer? *Applied Physiology, Nutrition, and Metabolism*, 39(6): 663–678. https://doi.org/10.1139/apnm-2013-0420.

Morales P, J., Valenzuela B, R., González M, D., González E, M., Tapia O, G., Sanhueza C, J. and Valenzuela B, A. (2012). Nuevas fuentes dietarias de acido alfa-linolénico: una visión crítica. [New dietary sources of alpha-linolenic acid: a critical view.] *Revista Chilena de Nutrición*, 39(3): 79–87. https://doi.org/10.4067/S0717-75182012000300012.

Morris, M. C., Evans, D. A., Bienias, J. L., Tangney, C. C., Bennett, D. A., Wilson, R. S. et al., (2003). Consumption of Fish and n-3 Fatty Acids and Risk of Incident Alzheimer Disease. *Archives of Neurology*, 60(7): 940. https://doi.org/10.1001/archneur.60.7.940.

Murphy, R. A., Mourtzakis, M., Mazurak, V. C. (2012). n-3 polyunsaturated fatty acids: the potential role for supplementation in cancer. *Curr Opin Clin Nutr Metab Care*, 15:246–51.

Neff, L. M., Culiner, J., Cunningham-Rundles, S., Seidman, C., Meehan, D., Maturi, J. et al., (2011). Algal docosahexaenoic acid affects plasma lipoprotein particle size distribution in overweight and obese adults. *The Journal of Nutrition*, 141(2): 207–213. https://doi.org/10.3945/jn.110.130021.

Norziah, M. H. and Ching, C. Y. (2000). Nutritional composition of edible seaweed Gracilaria changgi. *Food Chemistry*, 68(1): 69–76. https://doi.org/10.1016/S0308-8146(99)00161-2.

Oh, D. Y., Talukdar, S., Bae, E. J., Imamura, T., Morinaga, H., et al., (2010). GPR120 is an omega-3 fatty acid receptor mediating potent anti-inflammatory and insulin-sensitizing effects. *Cell*, 142:687–98.

Pawlosky, R. J., Hibbeln, J. R., Novotny, J. A. and Salem, N. (2001). Physiological compartmental analysis of alpha-linolenic acid metabolism in adult humans. *Journal of Lipid Research*, 42(8): 1257–1265.

Petrie, J. R., Shrestha, P., Zhou, X.-R., Mansour, M. P., Liu, Q., Belide, S. et al., (2012). Metabolic Engineering Plant Seeds with Fish Oil-Like

Levels of DHA. *PLoS ONE*, 7(11): e49165. https://doi.org/10.1371/journal.pone.0049165.

Ramprasath, V. R., Eyal, I., Zchut, S. and Jones, P. J. H. (2013). Enhanced increase of omega-3 index in healthy individuals with response to 4-week n-3 fatty acid supplementation from krill oil versus fish oil. *Lipids in Health and Disease*, 12: 178. https://doi.org/10.1186/1476-511X-12-178.

Rossoll, D., Bermúdez, R., Hauss, H., Schulz, K. G., Riebesell, U., Sommer, U. and Winder, M. (2012). Ocean acidification-induced food quality deterioration constrains trophic transfer. *PloS One*, 7(4): e34737. https://doi.org/10.1371/journal.pone.0034737.

Ruiz-Lopez, N., Haslam, R. P., Napier, J. A. and Sayanova, O. (2014). Successful high-level accumulation of fish oil omega-3 long-chain polyunsaturated fatty acids in a transgenic oilseed crop. *The Plant Journal*, 77(2): 198–208. https://doi.org/10.1111/tpj.12378.

Sala-Vila, A., Calder, P. C. (2011). Update on the relationship of fish intake with prostate, breast, and colorectal cancers. *Crit Rev Food Sci Nutr,* 51:855–71.

Sanz París, A., Sanchis, A. M., Malpartida, K. G. and García Gómez, D. C. (2012). Propuesta de perfil de ácidos grasos omega 3 en nutrición enteral. *Nutr Hosp. Nutr Hosp*, 2727(6). https://doi.org/10.3305/nh.2012.27.6.6023.

Schuchardt, J. P. and Hahn, A. (2013). Bioavailability of long-chain omega-3 fatty acids. *Prostaglandins, Leukotrienes and Essential Fatty Acids (PLEFA)*, 89(1): 1–8. https://doi.org/10.1016/J.PLEFA.2013.03.010.

Schuchardt, J. P., Schneider, I., Meyer, H., Neubronner, J., von Schacky, C. and Hahn, A. (2011). Incorporation of EPA and DHA into plasma phospholipids in response to different omega-3 fatty acid formulations--a comparative bioavailability study of fish oil vs. krill oil. *Lipids in Health and Disease*, 10: 145. https://doi.org/10.1186/1476-511X-10-145.

Schwab, U. S., C. Callaway, J., Erkkilä, A. T., Gynther, J., Uusitupa, M. I. J. and Järvinen, T. (2006). Effects of hempseed and flaxseed oils on the profile of serum lipids, serum total and lipoprotein lipid concentrations and haemostatic factors. *European Journal of Nutrition*, 45(8): 470–477. https://doi.org/10.1007/s00394-006-0621-z.

Serhan, C. N., Chiang, N., Van Dyke, T. E. (2008). Resolving inflammation: dual anti-inflammatory and pro-resolution lipid mediators. *Nat Rev Immunol*, 8:349–61.

Siebenlist, U., Franzoso, G. and Brown, K. (1994). Structure, Regulation and Function of NF-kappaB. *Annual Review of Cell Biology*, 10(1): 405–455. https://doi.org/10.1146/annurev.cb.10.110194.002201.

Sukenik, A. (1991). Ecophysiological considerations in the optimization of eicosapentaenoic acid production by Nannochloropsis sp. (Eustigmatophyceae). *Bioresource Technology*, 35(3): 263–269. https://doi.org/10.1016/0960-8524(91)90123-2.

Teng, Y. H., Aquino, R. S., Park, P. W. (2012). Molecular functions of syndecan-1 in disease. *Matrix Biol,* 31:3–16.

Thompson, L. U., Chen, J. M., Li, T., Strasser-Weippl, K. and Goss, P. E. (2005). Dietary Flaxseed Alters Tumor Biological Markers in Postmenopausal Breast Cancer. *Clinical Cancer Research*, 11(10): 3828–3835. https://doi.org/10.1158/1078-0432.CCR-04-2326.

Tilley, S. L., Coffman, T. M., Koller, B. H. (2001). Mixed messages: modulation of inflammation and immune responses by prostaglandins and thromboxanes. *J Clin Invest*, 108: 15–23.

Tocher, D. (2009). Issues surrounding fish as a source of omega-3 long-chain polyunsaturated fatty acids. *Lipid Technology*, 21(1): 13–16. https://doi.org/10.1002/lite.200800079.

Tocher, D. R. (2015). Omega-3 long-chain polyunsaturated fatty acids and aquaculture in perspective. *Aquaculture*, 449: 94–107. https://doi.org/10.1016/J.AQUACULTURE.2015.01.010.

Todorcević, M., Vegusdal, A., Gjøen, T., Sundvold, H., Torstensen, B. E., Kjaer, M. A. and Ruyter, B. (2008). Changes in fatty acids metabolism during differentiation of Atlantic salmon preadipocytes; effects of n-3 and n-9 fatty acids. *Biochimica et Biophysica Acta*, 1781(6–7): 326–335. https://doi.org/10.1016/j.bbalip.2008.04.014.

Truan, J. S., Chen, J.-M. and Thompson, L. U. (2010). Flaxseed oil reduces the growth of human breast tumors (MCF-7) at high levels of circulating estrogen. *Molecular Nutrition & Food Research*, 54(10): 1414–1421. https://doi.org/10.1002/mnfr.200900521.

Tu, W. C., Cook-Johnson, R. J., James, M. J., Mühlhäusler, B. S. and Gibson, R. A. (2010). Omega-3 long chain fatty acid synthesis is regulated more by substrate levels than gene expression. *Prostaglandins, Leukotrienes and Essential Fatty Acids*, 83(2): 61–68. https://doi.org/10.1016/j.plefa.2010.04.001.

Ulven, S. M. and Holven, K. B. (2015). Comparison of bioavailability of krill oil versus fish oil and health effect. *Vascular Health and Risk Management*, 11: 511. https://doi.org/10.2147/VHRM.S85165.

Ulven, S. M., Kirkhus, B., Lamglait, A., Basu, S., Elind, E., Haider, T. et al., (2011). Metabolic effects of krill oil are essentially similar to those of fish oil but at lower dose of EPA and DHA, in healthy volunteers. *Lipids*, 46(1): 37–46. https://doi.org/10.1007/s11745-010-3490-4.

Valenzuela, A. and Valenzuela, R. (2014). Acidos grasos omega-3 en la nutrición ¿como aportarlos? [Omega-3 fatty acids in nutrition, how to provide them?] *Revista Chilena de Nutrición*, 41(2): 205–211. https://doi.org/10.4067/S0717-75182014000200012.

Valenzuela, R., Tapia, G., González, M. and Valenzuela, A. (2011). Ácidos grasos omega-3 (EPA y DHA) y su aplicación en diversas situaciones clínicas. [Omega-3 fatty acids (EPA and DHA) and their application in various clinical situations.] *Rev Chil Nutr*, 38.

In: Polyunsaturated Fatty Acids (PUFAs)
Editor: Angel Catala
ISBN: 978-1-53613-572-5

Chapter 8

PROTEOMICS APPROACHES TO UNDERSTAND THE HEALTH EFFECTS OF MARINE OMEGA-3 PUFAS

Lucía Méndez and Isabel Medina*
Instituto de Investigaciones Marinas del CSIC (IIM-CSIC),
Vigo, Spain

ABSTRACT

Marine ω-3 polyunsaturated fatty acids (PUFAs), i.e., eicosapentaenoic acid (EPA) and docosahexaenoic acid (DHA), have demonstrated their bioactivity in the improvement of human health, by preventing or even palliating numerous diseases, including diet-related diseases such as cardiovascular diseases (CVD) and type 2 diabetes, but also neurogenerative diseases. Proteomics, which is defined as the large-scale analysis of proteins, offers valuable information about structure, function, post-translational modifications, interactions and abundance of proteins, information that has contributed to explain mechanisms of this bioactivity of marine ω-3 PUFAs. The aim of this chapter is to present an overview of the diverse information available in the bibliography that the

* Corresponding Author: medina@iim.csic.es.

application of proteomics approaches has provided to understand critical issues regarding bioactivity and functional properties of marine EPA and DHA in human health. Therefore, this chapter summarizes the proteomics tools adopted to test the beneficial effects of EPA and DHA consumption, the disease studied and the main findings resulting from Proteomics. This information highlights the usefulness of Proteomics for studying the beneficial effects of marine ω-3 PUFAs consumption and can guide researchers to better design proteomics approaches to test the bioactivity of these marine PUFAs.

1. INTRODUCTION

Marine omega-3 long-chain polyunsaturated fatty acids (PUFAs), especially eicosapentaenoic (EPA; 20:5, ω-3) and docosahexaenoic (DHA; 22:6, ω-3) have largely proved their bioactivity in human health (Venugopal and Shahidi, 1996). In general terms, bioactive compounds can be described as "essential and non-essential compounds (e.g., vitamins or polyphenols) that occur in nature, are part of the food chain, and can lead an effect on human health" (Biesalski et al. 2009). Therefore, bioactive compounds are natural food constituents that provide health benefits beyond their basic nutritional value. Their presence in a given foodstuff determines its classification as a functional food. Bioactive compounds are called nutraceuticals as well, a term coined in 1979 by Stephan DeFelice which results from the union of the terms "nutrition" and "pharmaceutical," which reflects both their presence in the human diet and their biological activity (Brower, 1998).

The bioactivity of marine long-chain PUFA lipids was firstly reported during the 1960s and 1970s when Bang and Dyerberg observed the low incidence of cardiovascular diseases and diabetes mellitus among Greenland Eskimo population, despite the high dietary fat content (Bang et al. 1976; Bang et al. 1971). These studies found that Greenlanders had fatty diets substantially enriched in marine long-chain ω-3 PUFAs because of the intake of high amounts of fish and marine mammalians. Later, several pieces of information collected from different experiments in animal models and clinical interventions have evidenced the anti-inflammatory properties of marine long-chain ω-3 PUFAs and their potential usefulness against autoimmune diseases. Additionally, their role on neurodegenerative diseases, cancer and several chronic diseases such as coronary heart disease, diabetes, obesity and insulin resistance have been supported as well (Simopoulos, 2002).

Nutrients alter metabolic functions in cells in a complex way. In fact, all cellular processes can be affected to some extend by diet composition. Thus, to understand properly the effect of diet on health it is necessary to study the mechanisms of nutrients and other bioactive food constituents at molecular level. As consequence, food science has been lately focused on developing technological tools sensitive enough to test the ability of seafood to prevent or correct several diseases and study how marine components affect the balance between health and disease. The new-omics methodologies (namely, genomics, transcriptomics, proteomics and metabolomics) that allow a large-scale analysis of cellular components by combining mass spectrometry techniques, bioinformatics or chemometrics, are postulated as promising tools to deal with food science demands (Moore and Weeks, 2011).

Specifically, proteomics tools are emerging as an important part of this holistic approach because proteins are important mediators of biological activities in all living cellular units. The term "proteomics" was firstly coined in 1994 by Marc Wilkins to refer to protein expression in a holistic manner (Wilkins and Appel, 2007). Nowadays, proteomics has extended its definition to "the large-scale analysis of the proteome" and therefore encompasses the methodologies used for the study of a proteome. It covers from protein identification and quantification up to the analysis of post-translational modifications, protein-protein interactions and protein function (Pandey and Mann, 2000). The "proteome" is comprised of all expressed proteins encoded by the genome of a cellular system, including all cellular proteins and all protein species (isoforms and protein modifications) (Wilkins and Gooley, 1997). Progress on proteomics has been technology-driven by a combination of conceptual breakthroughs and technical advances in separations techniques, mass spectrometry, protein chemistry and bioinformatics, which have transformed protein chemistry into the discipline of proteomics (Wilkins and Appel, 2007). Recently, Proteomics has emerged as a reliable and sensitive analytical methodology to monitor specific issues that endanger quality and safety of fishery products. It has applied to understand the mechanistic basis behind the beneficial effect of fish consumption to maintain the *in vivo* homeostatic balance and prevent diseases.

In spite of this consideration, genomics and transcriptomics tools continue to be the routine analysis to test the effect of food on life system. Only a few publications in nutritional research have applied the rest of omics in their studies (Moore and Weeks, 2011). In the field of marine food science, the application of proteomics approaches is markedly scarce and mainly related to cancer or chronic diet-induced diseases, such as obesity or cardiovascular diseases. The main seafood bioactive compounds studied have been the long chain omega-3 fatty acids. These compounds, especially EPA and DHA, which are highly concentrated on fishery products, have an important role on the nutritional quality of fish.

There are limited studies addressing the influence of marine lipids on proteome to illustrate their effects against human metabolic disorders. These studies have mainly used classical two-dimensional gel electrophoresis (2-DE) combined with mass spectrometry (MS) to elucidate changes in metabolic pathways and identify target proteins which may explain some of their effects on human health. This traditional methodology has made it possible to identify changes in several metabolic pathways, such as glucose and fatty acid metabolisms, oxidative stress, antioxidant defense mechanisms and redox status. The study of less abundant proteins such as those participating in inflammatory pathways requires the application in nutrition research of more quantitative and sensitive methods, like multiple reaction monitoring (MRM) and multiplexed immunoassays. Both strategies might be useful for the evaluation and validation of newly discovered candidate biomarkers in human biofluids. The present chapter will give an overview of the diverse and distinctive information that proteomic strategies have provided to study the health effects of the consumption of marine omega-3 PUFAs.

2. Proteomics to Investigate the Health Effects of Marine Omega-3 PUFAs

2.1. Marine Omega-3 PUFAs in Cancer Research Assayed by Proteomics

According to the available literature, proteomics approaches are suitable tools to investigate the potential beneficial effect of marine omega-3 PUFAs intake against several human diseases. Some of these researchers investigated the potential use of marine PUFAs against some types of cancer. In the investigation of prostate cancer, Zhao et al. (2016) tested the effects of a fish oil containing high amounts of omega-3 fatty acids (40% EPA, 40% DHA and 10% other omega-3 PUFAs) as compared with an olive oil containing high amounts of omega-9 oleic acid on viability of prostate cancer cell lines. Quantitative proteomics tools, based on label-free quantification by LC-ESI-MS/MS and validation by Western blot and multiple monitoring reaction (MRM) analysis, demonstrated that fish oil supplementation inhibited prostate cancer cell survival by inducing cytotoxicity after short treatment and suppressing inflammation pathways and inhibiting the glycolytic pathway in prolonged treatment. Authors also validated three protein targets modulated by fish oil (sequestosome-1, prostrate-associated microseminoprotein and sterol carrier protein 2) which suggested an additional mechanism involved on suppressing cell viability independently of the down-regulation of the fatty acid synthase. By two-dimensional electrophoresis (2-DE) gels and identification of proteins differently regulated by MALDI-TOF-MS/MS, Narayanan et al. (2006) evaluated the modulation of proteins in rat prostate cancer cells after the treatment with the combination of low doses of DHA and a chemopreventive agent (celecoxib). This treatment altered twelve proteins as compared with the control one. Among those, DHA and celecoxib significantly increased proteins involved in protein folding and synthesis, such as dnaK-type molecular chaperone grp78 (GRP78), heat shock 70 kDa proteins 5 and 8 (HSP70) and protein disulfide isomerase precursor (PDI).

Protein separation and quantification on 2-DE gels follow by identification with MALDI-TOF-MS/MS were also the chosen proteomics strategy to evaluate the role of marine omega-3 PUFAs in preventing lung cancer induced by tobacco carcinogens. Following this strategy, Chang et al. (2007) found that the intake of a high fish oil diet in rats treated with a potent lung carcinogen (nicotine-derived nitrosamine 4-(methylnitrosamino)-1-(3-pyridyl)-1-butanone) resulted in decreasing levels of apolipoprotein A-I and Clara cell 17-kDa, proteins which can constitute potential biomarkers of lung cancer induced by tobacco carcinogens.

The effects of marine omega-3 PUFAs against breast cancer were also assayed by applying proteomics approaches. By 2-DE and MALDI-TOF-MS/MS, Blanckaert et al. (2015) investigated the role of plasma membrane proteins in the DHA-induced inhibition of breast cancer cell invasion. DHA increased the levels of three membrane proteins (keratin, type II cytoskeletal 1 (KRT1), catalase and lamin-A/C) in the triple-negative breast cancer cell line MDA-MB-231. The main alteration found was the upregulation of KRT1 by DHA. Interestingly, *in vitro* assays demonstrated that this protein was directly involved in the anti-invasive activity of DHA in breast cancer cells.

A relatively novel proteomics technology, i.e., reverse phase protein array, was employed to shed light on the variable effects of high intakes of omega-3 EPA and DHA in reducing risk of pre-menopausal breast cancer. In pre-menopausal women with no prior invasive breast cancer, Fabian et al. (2015) found that daily consumption of a mix of EPA (1860 mg) and DHA (1500 mg) for 6 months exerted favorable effects in the modulation of several proteins associated with mitogen signaling and cell-cycle arrest. These findings evidence a promising use of high-dose EPA and DHA ethyl esters for primary prevention of breast cancer.

2.2. Marine Omega-3 PUFAs in Chronic Diseases Research Assayed by Proteomics

Besides cancer research, proteomics was also used to investigate the influence of marine oils, mainly EPA and DHA, on the prevention or palliation of several chronic human diseases, especially those related with the consumption of unhealthy diets.

Among these chronic disorders, some authors have evaluated the modulation of tissular proteins after omega-3 PUFAs intake in patients suffering metabolic syndrome (MetS). The MetS is a compilation of risk factors associated with CVD and type 2 diabetes. These factors include dyslipidemia (decreased high-density lipoprotein (HDL) cholesterol levels and increased low-density lipoprotein (LDL) cholesterol and triglycerides levels), hyperglycemia, hypertension, insulin resistance and obesity. Inflammation and oxidative stress have also been closely related to MetS and derived diseases (Grundy, 2008).

Quantitative proteomics tools based on 2-DE gels and MALDI-TOF-MS/MS were successfully applied to investigate the proteome modulation exerted by the supplementation with a mix of EPA and DHA, in a ratio of 1.4:1, of patients suffering MetS. The acute intake of these marine omega-3 PUFAs (Camargo et al. 2013) altered the proteome of peripheral blood mononuclear cells (PBMCs) by modulating five proteins belonging to cell signalling and interaction, DNA repair, cellular assembly and organization, and cell morphology. Likewise, the consumption of these fatty acids for 12 weeks altered up to 17 proteins of PBMCs proteome (Rangel-Zuñiga et al. 2015) involved in immunological diseases and inflammatory response, down-regulating proteins directly related to oxidative stress, inflammation, endoplasmic reticulum stress and DNA repair. Additionally, marine PUFAs down-regulated three proteins involved in the glucose metabolism in subcutaneous white adipose tissue (Jiménez-Gómez, et al. 2014). Together, these results were correlated with lower systemic insulin resistance and improved insulin signalling in the MetS patients after omega-3 intake.

Modulation of liver proteome by marine omega-3 PUFAs in metabolic syndrome induced by high fat/sucrose diets were also analyzed by proteomics. Two quantitative proteomics approaches, i.e., two-dimensional difference gel electrophoresis (2D-DIGE) coupled to MALDI-TOF-MS/MS and the isobaric tags for relative and absolute quantitation (iTRAQ)-labeling coupled to nano-LC-MS/MS, were applied to characterize the modulation of liver proteome exerted by the long-term (24 weeks) supplementation with fish oil (EPA:DHA 1:1) of standard or high fat and sucrose diet-fed rats (Méndez et al. 2017). Firstly, this combined strategy demonstrated that the physiological action of fish oil was closely dependent on the background diet. Indeed, six proteins changed into standard diets versus 31 proteins found altered into high fat and sucrose diets. Secondly, the specific changes on liver proteome induced by fish oil, especially when was added to the high fat, high sucrose diet, comprised lower levels of enzymes involved in the lipogenic and glycolitic pathways, higher levels of enzymes from fatty acid beta-oxidation and insulin signaling and the amelioration of endoplasmic reticulum stress. These changes were accompanied by less protein oxidation and the improvement of some biochemical measurements. Ahmed et al. (2014) also described similar trends in the liver of C57BL/6 mice fed standard diets by using 2-DE-MALDI-TOF-MS/MS. After four months of dietary intervention, fish oil up-regulated eight proteins related to lipid, carbohydrate, protein and one-carbon metabolism and the citric acid cycle, highlighting the integrated regulatory action of metabolic pathways by dietary fish oil. This regulation was correlated with fewer plasma triglycerides and free fatty acid levels. Wrzesinski et al. (2013) further addressed the effects of fish oil intake on the liver mitochondrial subproteome in rats suffering metabolic syndrome. Through 2-DE-MALDI-TOF-MS/MS and LC-MS/MS analysis, authors found that the supplementation with 10% of fish oil of a high fat diet for 50 weeks altered the levels of 54 mitochondrial proteins associated with the oxidation of fatty acid and amino acid metabolisms and the increase of oxidative phosphorylation.

The potential utility of marine omega-3 in the prevention of coronary heart disease (CHD) was also addressed in several proteomics studies available in literature. De Roos et al. (2005) applied 2-DE-MALDI-TOF-MS/MS to study the modulation of the liver proteome by fish oil supplementation in an animal model for lipid metabolism and atherosclerosis, the APOE*3 Leiden transgenic mice, fed a diet high in saturated fat and cholesterol for 3 weeks. After dietary intervention, 44 proteins resulted altered by fish oil, which were mainly involved in glucose and lipid metabolisms, as well as oxidation and aging processes. These changes were correlated with lower plasma and liver cholesterol and triglycerides levels as well as fewer amounts of plasma free fatty acids and glucose and higher plasma insulin levels, revealing new insights into mechanisms by which these fish oils can regulate lipid metabolism and related pathways. Later, the same authors (de Roos et al. 2008) showed the effectiveness in humans of a daily dose of 3.5 g of fish oil for six weeks in decreasing 10 serum proteins which were modulators of inflammation and/or lipid and lipoprotein metabolism. These down-regulations triggered the formation of larger, more cholesterol-rich HDL2 particle which might imply that fish oil activated anti-inflammatory and lipid modulating mechanisms believed to impede the early onset of CHD. This specific effect of fish oil on HDL metabolism has later been investigated by Burillo et al. (2012) using 2D-DIGE. Experimental design consisted in comparing the proteome of HDL in healthy smoker volunteers before and after ω-3 PUFAs intake for five weeks. Marine ω-3 PUFAs increased seven proteins related to the antioxidant, anti-inflammatory and anti-atherosclerotic properties of HDL. Likewise, marine lipids decreased the levels of six proteins involved in the regulation of complement activation and acute phase response. Moreover, the change in the lipoprotein containing apoAI (LpAI) proteome suggested that the protein changes found might have improved the functionality of the particle.

Proteomics tools also evaluated the differential effect of EPA and DHA in the context of metabolic disorders induced by diet. 2DE-DIGE strategy was used by Kalupahana et al. (2010) to test the modulation of proteins in 3T3-L1 adipocytes exerted after the treatment with EPA, as compared with the omega-6 arachidonic acid. In cells treated with EPA, 19 proteins belong to carbohydrate and fatty acid metabolism and cellular metabolism including response to stress, were up-regulated. Moreover, EPA treatment decreased eight proteins, which were related to lipogenesis and other cellular metabolic processes such as cytoskeleton organization and biogenesis. 2-DE gels and LC-ESI-MS/MS analysis were successfully used to distinguish the effect of fish oil, or only DHA, on liver proteome against the changes induces by a high fat and high cholesterol diet in an animal model of atherosclerosis (apoE knockout mice) (Mavrommatis et al. 2010). Both DHA and fish oil modulated 35 liver proteins, which were implicated in lipoprotein metabolism and oxidative stress. Interestingly, authors also found four proteins differentially regulated by DHA or fish oil. Therefore, even if both fish oil and DHA can improve lipoprotein metabolism and oxidative stress, only the intervention with DHA reduced protein levels of the hepatic soluble epoxide hydrolase, an enzyme closely associated with the development of cardiovascular diseases. The same proteomics approach was used by Johnson et al. (2015) to highlight the potentially independent protective/reversal effect of dietary EPA or DHA against mitochondrial dysfunction in aging skeletal muscle, after diet supplementation of young or older C57BL/6 mice. After 10 weeks, the decay in mitochondrial function related to age was attenuated only after the dietary supplementation with EPA but not with DHA. Thirty-nine mitochondrial proteins changed between old control and old EPA-treated mice and 32 between old control and old DHA-treated mice. Only three proteins for EPA supplementation and seven for DHA were coincident with those proteins found to differ in young and old control mice. Consequently, results showed that neither EPA nor DHA attenuated the age-related drop in mitochondrial protein content, but the proteome changes indicated some common biological effects of EPA and DHA, for instance regarding anticoagulation and anti-inflammatory properties and reduced FXR/RXR activation. However, only EPA improved muscle protein quality, by decreasing mitochondrial protein carbamylation, a post-translational protein modification (PTM) that is driven by inflammation.

A large body of evidence has associated an increased level of protein oxidation with the development and progression of some human diseases, including cancer, diabetes or cardiovascular diseases. On the other hand, it is well known that the consumption of ω-3 PUFAs influence redox homeostasis (Richard et al. 2008) and both antioxidant and pro-oxidant properties have been described for these fatty acids. Therefore, the study of oxidative PTMs might be critical for understanding the mechanisms that govern the action of marine ω-3 PUFAs in these pathologies. In fact, novel proteomics approaches such as redox proteomics have gained much more attention in nutritional research in the last years. Redox proteomics is a novel subfield of proteomics aimed at identifying and quantifying redox-based changes within the proteome both in redox signaling and under oxidative stress conditions (Butterfield and Dalle-Donne, 2012). However, redox proteomics tools have been scarcely employed in the characterization of proteins altered by fish oil consumption, mainly due to methodological limitations.

One publication evaluated the effect of different EPA and DHA ratios (1:1, 2:1, and 1:2, respectively) on protein carbonylation from plasma and some tissues (kidney, skeletal muscle, and liver) in rats after receiving a weekly oral dose of marine PUFAs for 13 weeks. Soybean (rich in ω-6 linoleic acid (LA)) or linseed oil (rich in ω-3 ALA) were used as control oils. Méndez et al. (2013) employed a proteomics approach, consisting of the specific label of carbonyls moieties formed on protein with the reagent fluorescein-5-thiosemicarbazide (FTSC), visualization of FTSC-label proteins on 1-DE or 2-DE gels, and protein identification by LC-ESI-MS/MS, to quantify and identify carbonylated protein in the selected tissues. Results showed that the three fish oil ratios, especially the 1:1 EPA:DHA, selectively decreased the carbonylation level of albumin in plasma, actin in skeletal muscle and albumin, argininosuccinate synthetase and 3-alpha-hydroxysteroid dehydrogenase in liver. This protective effect was accompanied by the improvement of biochemical features. Then, 2DE-MS/MS strategy was followed by Jourmard-Cubizolles et al. (2013) to study the formation of 4-hydroxynonenal (4-HNE) protein adducts derived from PUFA peroxidation in the aortic proteome from atherosclerotic prone (LDLR−/−) mice fed an atherogenic diet supplemented with DHA for 20 weeks. The supplementation with DHA altered the protein amount of 19 aortic proteins, mostly related to glucose or lipid metabolism. Moreover, in the DHA-supplemented group, authors found a significant up-regulation of the superoxide dismutase, which suggested a role of DHA for improving vascular antioxidant defenses. Results from the quantification of proteins with 4-HNE adducts were in agreement with the up-regulation of the superoxide dismutase. In fact, DHA supplementation did not increase the oxidation level of the 4-HNE-modifed proteins.

A summary of the articles cited in this chapter is shown in Table 1.

Table 1. Studies found in the literature that have used proteomics tools to investigate the health effects of EPA and DHA

Marine ω-3 PUFAs	Disease	Proteomics approach	Target proteome	Reference
40% EPA, 40% DHA	Prostate cancer	Label-free quantification by LC-ESI-MS/MS Validation by Western blot and MRM	Prostate cancer cells	Zhao et al. (2016)
DHA and celecoxib	Prostate cancer	2-DE-MS/MS	Prostate cancer cells	Narayanan et al. (2006)
17% fish oil for 3 months	Lung cancer	2-DE-MS/MS	Rat lung	Chang et al. (2007)
DHA	Breast cancer	2-DE-MS/MS	Breast cancer cells	Blanckaert et al. (2015)
EPA (1860 mg) and DHA (1500 mg) for 6 months	Breast cancer	Reverse phase protein array	Human breast cancer cells	Fabian et al. (2015)
Marine ω-3 PUFAs	Disease	Proteomics approach	Target proteome	Reference
Acute intake of EPA/DHA (1.4:1)	MetS	2-DE-MS/MS	Human PBMCs	Camargo et al. (2013)
EPA/DHA (1.4:1) for 12 weeks	MetS	2-DE-MS/MS	Human PBMCs	Rangel-Zúñiga al. (2015)
EPA/DHA (1.4:1) for 12 weeks	MetS	2-DE-MS/MS	Human white adipose tissue	Jiménez-Gómez al.(2014)

EPA/DHA (1:1) for 28 weeks	Metabolic alterations induced by high sucrose/high fat diet	2-DIGE-MS/MS iTRAQ-nanoLC-MS/MS	Rat liver	Méndez et al. (2017)
EPA/DHA (1:1) for 4 months	Healthy mice	2-DE-MS/MS	Mice liver	Ahmed et al. (2014)
EPA/DHA (2:1) for 50 weeks	MetS	2-DE-MS/MS	Rat liver mitochondria	Wrzesinski et al. (2013)
EPA/DHA (2:1) for 3 weeks	CHD	2-DE-MS/MS	Mice liver	de Roos et al. (2005)
Marine ω-3 PUFAs	Disease	Proteomics approach	Target proteome	Reference
EPA/DHA (2:1) for 6 weeks	CHD	2-DE-MS/MS	Human serum	de Roos et al. (2008)
0.6 g/d EPA&DHA for 5 weeks	CHD	2-DIGE-MS/MS	Human HDL	Burillo et al. (2012)
EPA	Cell culture	2-DIGE-MS/MS	3T3-L1 adipocytes	Kalupahana et al. (2010)
EPA/DHA (1.4:1) or DHA for 2 weeks	apoE knockout mice fed HFC diet	2-DE-MS/MS	Mice liver	Mavrommatis et al. (2010)
0.5% EPA or 0.5% DHA for 10 weeks	Ageing	Quantitative untargeted nanoLC-MS/MS	Mice quadriceps muscle	Johnson et al. (2015)
EPA:DHA 1:1 or 2:1 or 1:2 for 13 weeks	Healthy rats	FTSC-carbonyl protein labeling 1DE- and 2DE-MS/MS	Rat plasma, kidney, skeletal muscle, and liver	Méndez et al. (2013)
2% DHA for 20 weeks	CHD	2DE-MS/MS	Mice aorte	Jourmard-Cubizolles et al. (2013)

FINAL REMARKS AND CONCLUSION

The study of the proteins and metabolic pathways modulated by dietary marine ω-3 PUFAs has significantly contributed to progress on the knowledge of the mechanisms behind their beneficial health effects. Proteomics approaches have contributed to model biological processes controlled by dietary marine lipids, as well as to discover novel biomarkers for health or disease. Indeed, proteomics tools have allowed significant advances in cancer research by the discovery of promising potential biomarkers of some types of cancer, including prostate (sequestosome-1, prostrate-associated microseminoprotein and sterol carrier protein 2), lung (apolipoprotein A-I and Clara cell 17-kDa) and breast (Keratin, type II cytoskeletal 1).

On one hand, the application of proteomics approaches have demonstrated some mechanisms involved in the physiological action of dietary marine lipids, which have been previously proposed by genomics and transcriptomics. On the other hand, these techniques have also pointed out new molecular pathways and proteins regulated by omega-3 fatty acids.

Among the metabolic effects of dietary ω-3 PUFAs, proteomics found that marine oil down-regulated proteins participating in lipogenesis and glycolysis while causing the up-regulation of proteins involved in fatty acid beta-oxidation in several animal tissues, including the adipose tissue, aorta and liver. This dual action may explain the improvement of lipid profiles and the lower accumulation of fat often found after dietary interventions with these PUFAs. Urea cycle and protein metabolism, especially proteasome system were also found under the control of omega-3 intake. These findings demonstrated that EPA and DHA affect insulin signaling. Additionally, proteomics reported a substantial up-regulation of the antioxidant system by inducing higher levels of antioxidant enzymes, ameliorating of endoplasmic oxidative stress and stimulating protein and DNA cellular system repair.

It is especially noteworthy that proteomics exposed that the intake of marine lipids controls both protein quantity and quality, since they are able to modulate the generation of oxidative PTMs on proteins, mainly carbonyl moieties. The control of these oxidative PTMs may be the key factor to describe deeply the mechanisms behind the benefits of the consumption of marine lipids in decreasing the risk of a high number of human diseases, which are characterized by high levels of oxidative stress. The impairment of normal redox homeostasis, and the consequent accumulation of oxidized biomolecules, has been linked to the onset and/or development of a great variety of diet-induced diseases. For instance, EPA and DHA reduced carbonylation of proteins related to the antioxidant system such as albumin or 3-α-hydroxysteroid dehydrogenase in plasma and liver, respectively. Proteomics findings in liver demonstrated that EPA and DHA improved ammonia detoxification by decreasing carbonylation level of argininosuccinate synthetase while increasing oxidation of aspartate aminotransferase. Finally, in skeletal muscle, marine PUFAs protected from cellular dysfunction by ameliorating actin carbonylation level.

To sum up, proteomics constitute an important strategy to uncover the biological mechanisms that govern the health effect of omega-3 marine lipids. The usefulness of EPA and DHA consumption to prevent and/or palliate several types of cancer and chronic diseases closely related to the intake of unhealthy diets were the most often processes studied by using proteomic tools. The findings of these researches provide the basis to identify proper biomarkers of disease, design high-specific drugs with minimal side effects and personalized nutraceuticals and functional foods. Moreover, the more detailed characterization of mechanisms of marine PUFAs action may facilitate the identification of other diseases in which omega-3 lipids can exert their beneficial effects. The integration of proteomics with other –omics such as metabolomics and new advances in MS-based tools undoubtedly will provide reliable information in order to elucidate these mechanisms.

ACKNOWLEDGMENTS

Spanish Ministry of Science and Innovation is acknowledged for grant AGL2013-49079-C2-1-R.

REFERENCES

Ahmed, A. A., Balogun, K. A., Bykova, N. V., Cheema, S. K. 2014. Novel regulatory roles of omega-3 fatty acids in metabolic pathways: A proteomics approach. *Nutr. Metab*, 11, 6.

Bang, H. O., Dyerberg, G. J., Hjorne, N., 1976. The composition of food consumed by Greenland Eskimos. *Acta Medica Scandinavica* 200, 69-73.

Bang, H. O., Dyerberg, J., Nielsen, A. B., 1971. Plasma lipid and lipoprotein pattern in Greenlandic west-coast Eskimos. *Lancet* 1, 1143-1146.

Biesalski, H. K., Dragsted, L. O., Elmadfa, I., Grossklaus, R., Muller, M., Schrenk, D., Walter, P., Weber, P., 2009. Bioactive compounds: definition and assessment of activity. *Nutrition* 25, 1202-1205.

Blanckaert, V., Kerviel, V., Lépinay, A., Joubert-Durigneux, V., Hondermarck, H., Chénais, B., 2015. Docosahexaenoic acid inhibits the invasion of MDA-MB-231 breast cancer cells through upregulation of cytokeratin-1. *International Journal of Oncology* 46, 2649-2655.

Brower, V., 1998. Nutraceuticals: poised for a healthy slice of the healthcare market? *Nat Biotechnol* 16, 728-731.

Burillo, E., Mateo-Gallego, R., Cenarro, A., Fiddyment, S., Bea, A. M., Jorge, I., Vázquez, J., Civeira, F. 2012. Beneficial effects of omega-3 fatty acids in the proteome of high-density lipoprotein proteome. *Lipids Health Dis.*, 11, 116.

Butterfield, D. A., Dalle-Donne, I., 2012. Redox Proteomics. *Antioxidants & Redox Signaling* 17, 1487-1489.

Camargo, A., Rangel-Zuniga, O. A., Pena-Orihuela, P., Marin, C., Perez-Martinez, P., Delgado-Lista, J., Gutierrez-Mariscal, F. M., Malagon, M. M., Roche, H. M., Tinahones, F. J., Perez-Jimenez, F., Lopez-Miranda, J., 2013. Postprandial changes in the proteome are modulated by dietary fat in patients with metabolic syndrome. *J Nutr Biochem* 24, 318-324.

Chang, S. I., El-Bayoumy, K., Sinha, I., Trushin, N., Stanley, B., Pittman, B., Prokopczyk, B., 2007. 4-(Methylnitrosamino)-I-(3-pyridyl)-1-

butanone enhances the expression of apolipoprotein A-I and Clara cell 17-kDa protein in the lung proteomes of rats fed a corn oil diet but not a fish oil diet. *Cancer Epidemiol Biomarkers Prev* 16, 228-235.

De Roos, B., Duivenvoorden, I., Rucklidge, G., Reid, M., Ross, K., Lamers, R. J., Voshol, P. J., Havekes, L.M., Teusink, B. 2005. Response of apolipoprotein E*3-Leiden transgenic mice to dietary fatty acids: Combining liver proteomics with physiological data. *FASEB J.,* 19, 813–815.

De Roos, B., Geelen, A., Ross, K., Rucklidge, G., Reid, M., Duncan, G., Caslake, M., Horgan, G., Brouwer, I. A. 2008. Identification of potential serum biomarkers of inflammation and lipid modulation that are altered by fish oil supplementation in healthy volunteers. *Proteomics,* 8, 1965–1974.

Fabian, C. J., Kimler, B. F., Phillips, T. A., Nydegger, J. L., Kreutzjans, A. L., Carlson, S. E., Hidaka, B. H., Metheny, T., Zalles, C. M., Mills, G. B., Powers, K. R., Sullivan, D. K., Petroff, B. K., Hensing, W. L., Fridley, B. L., Hursting, S. D., 2015. Modulation of breast cancer risk biomarkers by high-dose omega-3 fatty acids: Phase II pilot study in postmenopausal women. *Cancer Prevention Research* 8, 922-931.

Grundy, S. M. 2008. Metabolic syndrome pandemic. *Arterioscler. Thromb. Vasc. Biol.*, 28, 629–63.

Jiménez-Gómez, Y., Cruz-Teno, C., Rangel-Zúñiga, O. A., Peinado, J. R., Pérez-Martínez, P., Delgado-Lista, J., García-Ríos, A., Camargo, A., Vázquez-Martínez, R., Ortega-Bellido, M., et al. 2014. Effect of dietary fat modification on subcutaneous white adipose tissue insulin sensitivity in patients with metabolic syndrome. *Mol. Nutr. Food Res.,* 58, 2177–2188.

Johnson, M. L., Lalia, A. Z., Dasari, S., Pallauf, M., Fitch, M., Hellerstein, M. K., Lanza, I. R. 2015. Eicosapentaenoic acid but not docosahexaenoic acid restores skeletal muscle mitochondrial oxidative capacity in old mice. *Aging Cell*, 14, 734–743.

Joumard-Cubizolles, L., Gladine, C., Gérard, N., Chambon, C., Brachet, P., Comte, B., Mazur, A. 2013. Proteomic analysis of aorta of LDLR-/- mice given omega-3 fatty acids reveals modulation of energy metabolism and oxidative stress pathway. *Eur. J. Lipid Sci. Technol.*, 115, 1492–1498.

Kalupahana, N. S., Claycombe, K., Newman, S. J., Stewart, T., Siriwardhana, N., Matthan, N., Lichtenstein, A. H., Moustaid-Moussa, N. 2010. Eicosapentaenoic acid prevents and reverses insulin resistance in high-fat diet-induced obese mice via modulation of adipose tissue inflammation. *J. Nutr.*, 140, 1915–1922.

Mavrommatis, Y., Ross, K., Rucklidge, G., Reid, M., Duncan, G., Gordon, M.J., Thies, F., Sneddon, A., de Roos, B., 2010. Intervention with fish oil, but not with docosahexaenoic acid, results in lower levels of hepatic soluble epoxide hydrolase with time in apoE knockout mice. *British Journal of Nutrition* 103, 16-24.

Méndez, L., Pazos, M., Gallardo, J. M., Torres, J. L., Pérez-Jiménez, J., Nogués, R., Romeu, M., Medina, I., 2013. Reduced protein oxidation in Wistar rats supplemented with marine omega-3 PUFAs. *Free Radic. Biol. Med.* 55, 8-20.

Méndez, L., Ciordia, S., Fernández, M. S., Juárez, S., Ramos, A., Pazos, M., Gallardo, J. M., Torres, J. L., Nogués, M. R., Medina, I. 2017. Changes in liver proteins of rats fed standard and high-fat and sucrose diets induced by fish omega-3 PUFAs and their combination with grape polyphenols according to quantitative proteomics. *J. Nutr. Biochem.,* 41, 84–97.

Moore, J. B., Weeks, M. E., 2011. Proteomics and systems biology: current and future applications in the nutritional sciences. *Adv Nutr* 2, 355-364.

Narayanan, N. K., Narayanan, B. A., Bosland, M., Condon, M. S., Nargi, D., 2006. Docosahexaenoic acid in combination with celecoxib modulates HSP70 and p53 proteins in prostate cancer cells. *International Journal of Cancer* 119, 1586-1598.

Pandey, A., Mann, M., 2000. Proteomics to study genes and genomes. *Nature* 405, 837-846.

Rangel-Zuniga, O. A., Camargo, A., Marin, C., Pena-Orihuela, P., Perez-Martinez, P., Delgado-Lista, J., Gonzalez-Guardia, L., Yubero-Serrano, E. M., Tinahones, F. J., Malagon, M. M., Perez-Jimenez, F.,

Roche, H. M., Lopez-Miranda, J., 2015. Proteome from patients with metabolic syndrome is regulated by quantity and quality of dietary lipids. *BMC Genomics* 16, 509.

Richard, D., Kefi, K., Barbe, U., Bausero, P., Visioli, F. 2008. Polyunsaturated fatty acids as antioxidants. *Pharmacol. Res*., 57, 451–455.

Simopoulos, A. P., 2002. Omega-3 Fatty Acids in Inflammation and Autoimmune Diseases. *Journal of the American College of Nutrition* 21, 495-505.

Venugopal, V., Shahidi, F., 1996. Structure and composition of fish muscle. *Food Reviews International* 12, 175-197.

Wilkins, M., Gooley, A. 1997. Protein Identification in Proteome Projects. In *Proteome Research: New Frontiers in Functional Genomics*; Wilkins, M., Williams, K., Appel, R., Hochstrasser, D., Eds.; Springer: Berlin/Heidelberg, Germany; 35–64.

Wilkins, M. R., Appel, R. D., 2007. Ten Years of the Proteome, in: Wilkins, M. R., Appel, R. D., Williams, K. L., Hochstrasser, D. F. (Eds.), *Proteome Research: Concepts, Technology and Application*. Springer Berlin Heidelberg, Berlin, Heidelberg, pp. 1-13.

Wrzesinski, K., R. León, I., Kulej, K., Sprenger, R. R., Bjørndal, B., Christensen, B. J., Berge, R. K., N. Jensen, O., Rogowska-Wrzesinska, A., 2013. Proteomics identifies molecular networks affected by tetradecylthioacetic acid and fish oil supplemented diets. *Journal of Proteomics* 84, 61-77.

Zhao, H., Pflug, B. R., Lai, X., Wang, M., 2016. Metabolic and molecular regulation of dietary polyunsaturated fatty acids on prostate cancer. *Proteomics - Clinical Applications* 10, 267-279.

In: Polyunsaturated Fatty Acids (PUFAs) ISBN: 978-1-53613-572-5
Editor: Angel Catala

Chapter 9

ESSENTIAL FATTY ACIDS: HEALTH IMPLICATIONS AND THREATS TO SOURCES

H. A. Leaver[1,*] and D. Rotondo[2]
[1]University of Edinburgh, Edinburgh, UK
[2]Strathclyde Institute of Pharmacy & Biomedical Sciences, University of Strathclyde, Glasgow, UK

ABSTRACT

The importance of essential fatty acids (EFA) in supporting human health is well established. EFAs represent a structural and metabolic source of versatile agents necessary for membrane, cellular and trans-cellular function. While recommendations on n-3 and n-6 EFA in the diet have been defined by health regulatory bodies, population studies indicate that typical diets do not follow these recommendations. Insufficient dietary EFA contribute to many health problems such as cardiovascular disease and other morbidities. Current advances in biochemistry and biotechnology are improving knowledge of EFA activities and synthesis. Biochemical approaches may help to clarify the details of EFA interactions with metal ions, oxidative metabolites and transmembrane flux. Also, biotechnology production processes are improving the yield

*Corresponding author Hanne.leaver@gmail.com.

and sources of EFA, which, together with improved monitoring of fatty acids and their metabolites will allow improved understanding of dietary dynamics and effects on health. However, a serious problem is that the principal sources of EFA are threatened by environmental pollution. Without the intervention of current food production, processing and legislation to limit environmental pollution, it is likely that insufficient EFA supplies will be available to support a healthy global population.

Essential Fatty Acids in Health

The essential fatty acids (EFA) are defined in terms of their essential nature as a component of mammalian diet, as they are not synthesised in mammals, and must be derived from the diet [1]. Diverse dietary sources include marine and plant sources and meat and dairy products also represent a rich source of n-6 EFA. Dietary EFA deficiency in humans is rare, but recommended dietary EFA levels are not currently observed in most populations, with conflicting expert advice on diet, suboptimal EFA content of dietary staples and EFA content in dietary products being significant factors.

In terms of dietary EFA intake, dietary recommendations on n-3 and n-6 polyunsaturated fatty acids have been defined by health regulatory bodies [2]. The European Food Safety Authority (EFSA) recommends 4% of energy intake (en%) as linolenic acid (ALA) and 0.5 en% as alpha linolenic acid and 250 mg eicosapentaenoic acid (EPA) and docosahexaenoic acid (DHA), with additional DHA recommendations for infants [3]. Similarly, the World Health Organisation (WHO), recommends 6-11 en% n-6 fatty acids and 0.5-2 en% n-3 fatty acids [3]. However, population studies indicate that typical diets do not follow these recommendations [4], which in turn contribute to cardiovascular disease and other morbidities.

Epidemiological evidence of EFA requirements include dietary EFA deprivation studies in young people and adolescents, and infant nutrition studies [5]. Symptoms of deficiency include skin disorders, thrombocytopenia, and in children, cognitive impairment, neuropathy and visual problems. Analysis of the high n-3 content of the Eskimo diet and bleeding time experiments in the 1960s suggested specific roles for n-3 EFA in haemostasis and cardiovascular health [6].

Biochemical and clinical studies have shown that the EFA represent a structural and metabolic source of versatile agents necessary for membrane, cellular and trans-cellular function. Biochemical studies have been crucial in identifying the EFA and their metabolism. The pathways of EFA synthesis, together with functions of individual metabolites are shown in Figure 1. The EFA are an essential part of membrane structure, interacting with membrane bound receptors, transporters and supramolecular assemblies associated with cell signalling and cellular functions.

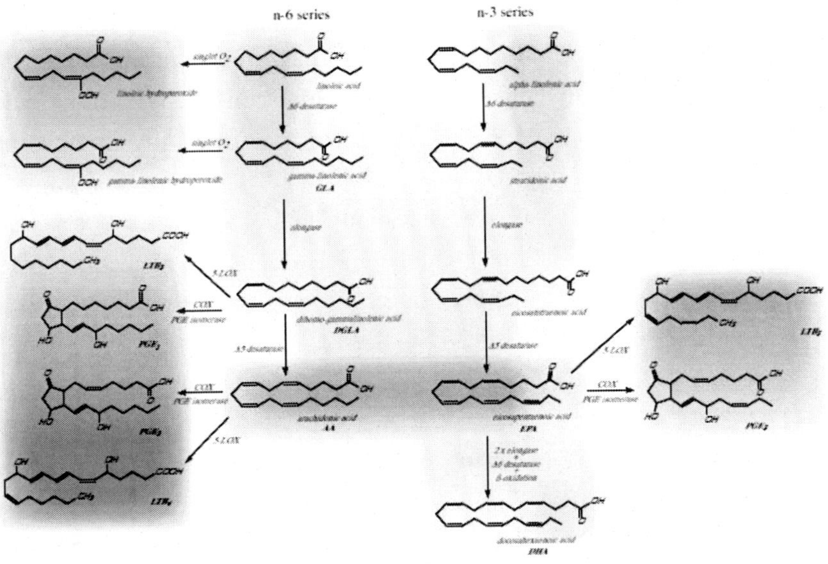

Figure 1. Functional aspects of essential fatty acid metabolism.

The pathways of n-3 n-6 EFA synthesis and eicosanoid production are highlighted in terms of their functions and dietary sources. Predominant dietary sources of EFA are shown in blue (C18 sources relate to both n-3 and n-6 EFA, but C22 sources are predominantly n-3 EFA of marine origin). EFA with a major structural role in membrane phospholipids are highlighted in yellow. EFA metabolites with intracellular signalling functions are indicated in red. EFA with both signalling and structural membrane functions are shown in orange. Enzymes of C20 EFA metabolism shown are cyclooxygenase (COX) and 5-lipoxygenase (5-LOX).

The importance of the essential fatty acids in supporting human health is generally recognised. Deficits of EFA have been detected in cardiovascular, oncological and psychiatric disease and treatments involving the modification of EFA metabolism which include some of the most widespread of pharmaceutical interventions, the non steroidal anti inflammatory agents. The health effects for which there is strong evidence of EFA are listed in Table 1. A basic biochemical property of EFA is its structural and functional role in cell membranes. The EFA are an integral component of cell membrane phospholipid structure, which is essential for compartmentation and flux of metabolites, secretory and bioenergetic functions.

EFA of the n-6 and n-3 series overlap in many physiological membrane functions, however, differences in their metabolism and signalling functions have led to dietary and pharmaceutical advances [2]. Partial dietary substitution of arachidonic acid by eicosapentaenoic acid may reduce the incidence of thrombotic disorders, including myocardial infarction [6]. Platelet lipid analysis demonstrated that high consumption of n-3 EFA by Eskimos was associated with increased n-3 EFA, together with functional changes. Eskimos had a significantly longer bleeding time, due to reduced platelet aggregation. Platelet C20:5 is converted by vascular tissue to anti-aggregatory prostacyclin. Beneficial effects of dietary n-3 EFA have also been shown in the secondary prevention of coronary heart disease, hypertension, and in some patients with diseases with an inflammatory component, including rheumatoid arthritis, ulcerative colitis and Crohn's disease [7]. As with other dietary manipulations, the proportion of EFA in the diet, together with dose and duration of EFA administration, are crucial. These manipulations aim to reverse changes in diet over the past 150 years, which have resulted in alterations in dietary EFA consumption, and increased n-6/n-3 EFA in cell membranes. There is also recent evidence that increased dietary n-6/n-3 ratios upregulate inflammatory mediator production, and inflammatory gene expression [8].

The skin epidermis has highly active EFA metabolism [9]. Dietary C18 n-6 EFA deficiency results in a characteristic scaly skin disorder and excessive epidermal water loss. In the epidermis, the C20 n-6 EFA arachidonic acid, is metabolized via the cyclooxygenase pathway into PGE_2, $PGF_{2\alpha}$ and PGD_2 and via the lipoxygenase pathway into 15-HETE [10]. Prostaglandins modulate normal skin physiological processes at low concentrations and inflammatory reactions at high concentrations. Similarly, the highly active epidermal 15-lipoxygenase enzymes metabolise n-3 EFA into monohydroxy acids with anti-inflammatory properties. In contrast, the C18 n-6 EFA is transformed into 13-HODE, which exerts antiproliferative properties in the tissue. Thus supplementation of diets with vegetable oils and/or fish oil or topical administration of these oils may promote the formation of local cutaneous metabolites for the management of skin inflammatory disorders [9]. A further example of n-6 dietary EFA playing a key role in physiological function is in the initiation of parturition, where dietary n-3 EFA inhibited prostaglandin production and parturition [11].

Table 1. Essential Fatty Acids in Human Health. Processes and conditions for which there is strong evidence of n-6 and n-3 EFA involvement as a primary factor. This evidence includes primary effects identified in dietary depletion of EFA, pharmacological inhibitor and biochemical studies.

Function	*physiological/pathological processes*	*Ref*
Cell membrane structure	permeability, transport	[12]
Skin function	psoriasis, eczema, acne	[9, 10]
Haemostasis	coagulation	[6]
Inflammation	inflammatory disease	[8]
Intestinal integrity	gastric lesions	[13]
Parturition	initiation	[11, 14]
Retinal function	visual acuity	[7]

Table 2. Essential Fatty Acids in Human Health. Processes and conditions for which there is evidence of n-6 and n-3 EFA involvement in pathophysiological aspects of function. These multi factorial pathophysiologies involve membrane and signalling pathways involving EFA. Some of these pathways overlap with functions defined by the dietary and pharmacological depletion studies described in Table 1, but others show evidence of processes and specific metabolites that are quite distinct. The evidence for a primary role of EFA in these processes is less defined and requires further investigation.

System	*Pathophysiological process/mediators*	*Ref*
Cancer	Proliferation, cell death	[15-17]
Renal function	Hypotension PGE, hyperplasia 20-HETE	[18]
Cardiovascular system	Vascular endothelium TXA_2, PGI_2, PGE_2	[19]
Immune system	Inflammatory processes PGE_2, LTs	[8, 20]
Pulmonary	Asthma LTs	[21-23]
Nervous system	Schizophrenia, bipolar disorder	[24-26]

Beneficial effects of n 3 fatty acids have been shown in the secondary prevention of type 2 diabetes, and, in some patients with renal disease and chronic obstructive pulmonary disease [7]. The role of eicosanoids in allergy and asthma has been well established [23]. Chronic endobronchial inflammation is a hallmark of pediatric asthma and involves the arachidonic acid pathway [22]. In the nervous system, EFA and bioactive lipids of both n-3 and n-6 pathways modulate the function of ion channels, receptors, neurotransmitter release, synaptic plasticity, glia–neuron relationships and neuronal gene expression. Thus, EFA signalling participates in cell function and in specific pathologic conditions of the nervous system. The availability of arachidonic acid is regulated by the biosynthesis of specific phospholipid molecular species, their EFA remodelling, and the regulation of the release of EFA from membrane reservoirs and the subsequent conversion steps including cyclooxygenation and lipoxygenation [24]. Additionally, the EFA including docosahexaenoic acid and arachidonic acid are suspected to play a key role in the pathogenesis of diabetes [27].

This review will analyse EFA health benefits in relation to current and potential supply, the importance of biochemical and pharmacological approaches to defining the role of EFA in health, and the serious problems that lack of international regulation and planning play in limiting global supplies of these essential nutrients.

BIOCHEMICAL APPROACHES TO DETERMINING HEALTH EFFECTS OF ESSENTIAL FATTY ACIDS

Biochemical approaches have provided the basis of current knowledge of EFA structure and metabolism. While analytical aspects of biochemical technology are widely used, less attention is often given to the potential of biochemical approaches in elucidating health aspects of EFA. Indeed, analytical approaches tend to overlook the significance of metabolic flexibility in sequestering and mobilising EFA at specific localities and under different pathophysiological conditions. However, together with cell biological and biophysical techniques, biochemical approaches to EFA synthesis, uptake and metabolism continue to clarify the details of EFA dynamics, storage and bioavailability. Thus, the role of membrane remodelling in sequestering EFA in key stores [15, 23] is being defined. Arachidonic acid remodeling through phospholipids impacts cell survival [23]. The regulation of AA/phospholipid-remodeling enzymes, cytosolic phospholipase $A_2\alpha$ and CoA-independent transacylase, provides a mechanism for altered eosinophil survival during allergic asthma.

Also, biochemical and dietary studies are defining specific roles of individual n-3 and n-6 EFA in pathophysiological processes. One of the better characterised systems is the inflammatory process in leukocytes. For example, four weeks of n-3 EFA diet reduces inflammatory processes gene in leukocytes, reducing monocyte PI3K, Akt, NFkappaB expression and inflammatory cytokines, and substantially (40%) reducing the n-6/n-3 EFA ratio in leukocyte membranes. The capacity of stimulated neutrophils to produce leukotriene B4 decreased by 31%. These data suggest that EFA can regulate signal transduction gene expression and proinflammatory cytokine gene transcription [8].

Similarly, the role of EFA in controlling genetic and epigenetic functions indicate new signalling functions [28]. Currently, fundamental biochemical advances include the study of EFA interactions with metal ions, oxidative metabolism, and EFA transmembrane flux. Further biochemical analysis of EFA sources, their uptake and storage and EFA metabolism is necessary in order to define more closely their role in physiological and pathological processes.

Nutritional, Nutraceutical and Pharmaceutical Aspects of Essential Fatty Acid Actions

The nutritional and pharmaceutical industries are prominent in EFA research and development. Nutritional studies into dietary polyunsaturated fatty acids have identified key functions of EFA of the n-3 and n-6 series. They have also indicated the advantages of dietary intake of EFA in combination with associated antioxidants, dietary cofactors and vitamins.

Nutritional recommendations have led to EFA supplementation and to nutraceutical developments. Current trends in nutritional research have led to increased dietary supplementation of the food and food chain with EFA [29]. While considerable research has been carried out into the stabilisation of EFA sources, both in nutritional supplies and supplements, less is known about the nutritional effects of these additives and their associated long term health risks. Thus, critical metabolic and genetic factors influence the conversion of C18 precursor EFA C20:3n-6 EFA, which consequently affect the balance of C20 derived metabolites [20]. These factors affect the clinical effectiveness of EFA supplements in treating inflammatory and other conditions. Additionally, other factors affecting efficacy include the opposing effects of n-3 and n-6 EFA and their metabolites on inflammatory processes and diseases and the impact of genetic variations within the fatty acid desaturase on n-3/n-6 ratios and bioactive metabolites. These factors influence the heterogeneity of results observed in EFA supplement-based clinical trials and suggest that individual approaches utilizing PUFA-based supplements may be appropriate for the prevention and treatment of complex human diseases [21, 29].

Pharmacological approaches include the modification of PUFA metabolism using anti inflammatory agents [30]. Pharmaceutical agents with anti inflammatory action of synthetic and natural origin continue to be developed with selective action on 5-lipoxygenase, attenuating leukotriene B4 production [31, 32]. In oncogenesis, multi factorial factors include P450 derived lipid oxidation products [33]. Endogenously produced lipid autacoids, such as the P450 metabolite of arachidonic acid 20-HETE, metabolite of arachidonic acid, to inhibit proliferation of human renal carcinoma are locally acting small molecule mediators that play a central role in the regulation of tissue homeostasis and influence tissue hyperplasia [18]. These represent novel targets for pharmaceutical and nutraceutical development.

THREATS TO ESSENTIAL FATTY ACID SUPPLIES AND HEALTH

A fundamental problem in EFA in health lies in dietary supply, which, together with lack of dietary education, customs, lifestyle and commercial food industry practices, currently limit the supply of adequate EFA in the diets currently ingested by most populations (Table 3). Thus, expert recommendations on EFA content and n-3/n-6 ratios in many processed foods are often ignored in commercial practice.

A further problem lies in the contamination of the principal sources of essential fatty acids by environmental pollutants, together with the lack of environmental planning to limits to this contamination [34]. Contamination of marine pufa is a major problem in EFA production and one that is getting worse, due to lack of international regulation or implementation of current regulations. Specific contaminants with associated toxicity include heavy metal complexes, mercury, persistent organic pollutants, such as the polychlorinated biphenyls, and the polybrominated diphenyl ethers, plastics, plasticisers including phthalates which leach into ecosystems [35]. Land borne toxins including pesticides and hormone disruptive agents, also represent a major problem, affecting plant supplies and pollination.

Table 3. Threats to essential fatty acids supplies and health

Problem Areas	*Specific problems relating to EFA*
Health education	Lack of EFA knowledge in the general population
Food production/processing	EFA lack in dietary staples
Legislation	Lack of directives in food production and processing
Planning	Lack of promotion of EFA enrichment in dietary sources
Environmental degradation	Pollution, contaminated EFA supplies/limited EFA production
Biotechnology	Lack of EFA centered planning and implementation

Biotechnology approaches are improving the yield and sources of essential fatty acids, both in marine and terrestrial sources. These advances in production, together with improved monitoring of fatty acids and their metabolites, allow dietary dynamics and effects on health. However, there are considerable commercial confounding pressures in biotechnology, due to the introduction of practices such as pesticide dependence and terminator genetic manipulation, which have led to public concern about the environmental effects of genetic modification without closer regulation.

In conclusion, without the intervention of improvements in current food production and processing, and legislation to limit environmental pollution, it is likely that insufficient essential fatty acid supplies will be available to support a healthy global population. It is important that advances in biochemical, dietary and pharmaceutical EFA research are used to inform both the food production industry and the general public.

ACKNOWLEDGMENT

We are very grateful to Giuliana Rotondo for the preparation of Figure 1.

REFERENCES

[1] Holub, D. J. and B. J. Holub, Omega-3 fatty acids from fish oils and cardiovascular disease. *Mol. Cell Biochem.*, 2004. 263(1-2): p. 217-25.

[2] Wahle, K. W. J., S. D. Heys, and D. Rotondo, Authorised EU health claims for the essential fatty acids: n-6 linoleic (18:2n-6) and n-3 α-linolenic (18:3n-3) acids, in *Foods, Nutrients and Food Ingredients with Authorised EU Health Claims Volume 2*, M. J. Sadler, Editor. 2015, Woodhead Publishing: Elsevier - https://doi.org/10.1016/B978-1-78242-382-9.09987-4.

[3] WHO, Fats and Fatty Acids in Human Nutrition: Joint FAO/WHO Expert Consultation, Geneva, November 2008. *Annals of Nutrition and Metabolism*, 2009. 55(1-3): p. 1-273.

[4] Forsyth, S., S. Gautier, and N. Salem, Jr., Global Estimates of Dietary Intake of Docosahexaenoic Acid and Arachidonic Acid in Developing and Developed Countries. *Ann. Nutr. Metab.*, 2016. 68(4): p. 258-67.

[5] Decsi, T., D. Molnar, and B. Koletzko, The effect of under- and overnutrition on essential fatty acid metabolism in childhood. *Eur. J. Clin. Nutr.*, 1998. 52(8): p. 541-8.

[6] Dyerberg, J. and H.O. Bang, Haemostatic function and platelet polyunsaturated fatty acids in Eskimos. *Lancet,* 1979. 314(8140): p. 433-5.

[7] Simopoulos, A. P., The importance of the ratio of omega-6/omega-3 essential fatty acids. *Biomed. Pharmacother.*, 2002. 56(8): p. 365-79.

[8] Weaver, K. L., P. Ivester, M. Seeds, L. D. Case, J. P. Arm, and F. H. Chilton, Effect of dietary fatty acids on inflammatory gene expression in healthy humans. *J. Biol. Chem.,* 2009. 284(23): p. 15400-7.

[9] Ziboh, V. A., *Nutritional Modulation of Inflammation by Polyunsaturated Fatty Acids/Eicosanoids, in Nutrition and Immunology*, G. M. E., G. J. B., and K. C. L., Editors. 2000, Humana Press: New Jersey.

[10] Horrobin, D. F., Essential fatty acids in clinical dermatology. *J. Am. Acad. Dermatol.*, 1989. 20(6): p. 1045-53.

[11] Leaver, H. A., A. Howie, N. H. Wilson, P. L. Yap, and I. D. Aitken, Developmental changes in the fatty acids of rat uterus and the influence of dietary essential fatty acids. *Prostaglandins Leukot. Essent. Fatty Acids*, 1992. 46(2): p. 123-31.

[12] Gunstone, F. D. and F. A. Norris, *Lipids in Foods: Chemistry, Biochemistry and Technology*. 2013, London: Pergamon.

[13] Brzozowski, T., P. C. Konturek, S. J. Konturek, Z. Sliwowski, D. Drozdowicz, J. Stachura, R. Pajdo, and E. G. Hahn, Role of prostaglandins generated by cyclooxygenase-1 and cyclooxygenase-2 in healing of ischemia-reperfusion-induced gastric lesions. *Eur. J. Pharmacol.*, 1999. 385(1): p. 47-61.

[14] Leat, W. M. and C. A. Northrop, Effect of dietary linoleic and linolenic acids on gestation and parturition in the rat. *Q. J. Exp. Physiol.*, 1981. 66(2): p. 99-103.

[15] Chilton, F. H., R. C. Murphy, B. A. Wilson, S. Sergeant, H. Ainsworth, M. C. Seeds, and R. A. Mathias, Diet-gene interactions

and PUFA metabolism: a potential contributor to health disparities and human diseases. *Nutrients*, 2014. 6(5): p. 1993-2022.

[16] Monjazeb, A. M., K. P. High, A. Connoy, L.S. Hart, C. Koumenis, and F.H. Chilton, Arachidonic acid-induced gene expression in colon cancer cells. *Carcinogenesis*, 2006. 27(10): p. 1950-60.

[17] Rizzo, M. T., Cyclooxygenase-2 in oncogenesis. *Clin. Chim. Acta.*, 2011. 412(9-10): p. 671-87.

[18] Alexanian, A., V. A. Rufanova, B. Miller, A. Flasch, R. J. Roman, and A. Sorokin, Down-regulation of 20-HETE synthesis and signaling inhibits renal adenocarcinoma cell proliferation and tumor growth. *Anticancer Res*., 2009. 29(10): p. 3819-24.

[19] Soma, M., M. S. Manku, D. K. Jenkins, Y. S. Huang, and D. F. Horrobin, Effects of dietary saturated, N-6 and N-3 polyunsaturated fats on blood pressure and prostaglandins outflow from perfused mesenteric vascular bed in rats. *Prostaglandins*, 1985. 30(6): p. 969-80.

[20] Sergeant, S., E. Rahbar, and F. H. Chilton, Gamma-linolenic acid, Dihommo-gamma linolenic, Eicosanoids and Inflammatory Processes. *Eur. J. Pharmacol*., 2016. 785: p. 77-86.

[21] Arm, J. P., J. A. Boyce, L. Wang, H. Chhay, M. Zahid, V. Patil, U. Govindarajulu, P. Ivester, K. L. Weaver, S. Sergeant, E. Israel, and F.H. Chilton, Impact of botanical oils on polyunsaturated fatty acid metabolism and leukotriene generation in mild asthmatics. *Lipids Health Dis.*, 2013. 12: p. 141.

[22] Glowacka, E., U. Jedynak-Wasowicz, M. Sanak, and G. Lis, Exhaled eicosanoid profiles in children with atopic asthma and healthy controls. *Pediatr. Pulmonol*., 2013. 48(4): p. 324-35.

[23] Seeds, M. C., K. K. Peachman, D. L. Bowton, K. L. Sivertson, and F.H. Chilton, Regulation of arachidonate remodeling enzymes impacts eosinophil survival during allergic asthma. *Am. J. Respir. Cell Mol. Biol.*, 2009. 41(3): p. 358-66.

[24] Bazan, N. G., *Arachidonic Acid Signalling in the Nervous System*, in *eLS*. 2006, John Wiley & Sons, Ltd: online DOI:10.1038/npg.els. 0004045.

[25] Imig, J. D., A. N. Simpkins, M. Renic, and D. R. Harder, Cytochrome P450 eicosanoids and cerebral vascular function. *Expert Rev. Mol. Med.*, 2011. 13: p. e7.

[26] Sumiyoshi, T., Y. Higuchi, M. Matsui, H. Itoh, T. Uehara, T. Itoh, H. Arai, C. Takamiya, M. Suzuki, and M. Kurachi, Membrane fatty acid levels as a predictor of treatment response in chronic schizophrenia. *Psychiatry Res.,* 2011. 186(1): p. 23-7.

[27] Koehrer, P., S. Saab, O. Berdeaux, R. Isaico, S. Gregoire, S. Cabaret, A.M. Bron, C.P. Creuzot-Garcher, L. Bretillon, and N. Acar, Erythrocyte phospholipid and polyunsaturated fatty acid composition in diabetic retinopathy. *PLoS One*, 2014. 9(9): p. e106912.

[28] Rahbar, E., H. C. Ainsworth, T. D. Howard, G. A. Hawkins, I. Ruczinski, R. Mathias, M.C. Seeds, S. Sergeant, J. E. Hixson, D. M. Herrington, C. D. Langefeld, and F. H. Chilton, Uncovering the DNA methylation landscape in key regulatory regions within the FADS cluster. *PLoS One*, 2017. 12(9): p. e0180903.

[29] Chilton, F. H., R. Dutta, L. M. Reynolds, S. Sergeant, R. A. Mathias, and M. C. Seeds, Precision Nutrition and Omega-3 Polyunsaturated Fatty Acids: A Case for Personalized Supplementation Approaches for the Prevention and Management of Human Diseases. *Nutrients*, 2017. 9(11).

[30] Khanapure, S. P., D. S. Garvey, D. R. Janero, and L. G. Letts, Eicosanoids in inflammation: biosynthesis, pharmacology, and therapeutic frontiers. *Curr. Top Med. Chem.*, 2007. 7(3): p. 311-40.

[31] Griswold, D. E., P. J. Marshall, J. C. Lee, E. F. Webb, L. M. Hillegass, J. Wartell, J. Newton, Jr., and N. Hanna, Pharmacology of the pyrroloimidazole, SK&F 105809--II. Antiinflammatory activity and inhibition of mediator production in vivo. *Biochem. Pharmacol.,* 1991. 42(4): p. 825-31.

[32] Zeng, H., X. Liu, S. Dou, W. Xu, N. Li, X. Liu, W. Zhang, Z. Hu, and R. Liu, Huang-Lian-Jie-Du-Tang exerts anti-inflammatory effects in rats through inhibition of nitric oxide production and eicosanoid biosynthesis via the lipoxygenase pathway. *J. Pharm. Pharmacol.*, 2009. 61(12): p. 1699-707.

[33] Panigrahy, D., A. Kaipainen, E. R. Greene, and S. Huang, Cytochrome P450-derived eicosanoids: the neglected pathway in cancer. *Cancer Metastasis Rev.*, 2010. 29(4): p. 723-35.

[34] Monbiot, G., *How Did We Get Into This Mess?: Politics, Equality, Nature*. 2016, London: Verso.

[35] WHO, *Persistent Organic Pollutants: Impact on Child Health*. 2010: WHO.

ABOUT THE EDITOR

Prof. Dr. Angel Catalá
Universidad Nacional de La Plata (UNLP),
National Research Council(CONICET), Argentina

Prof. Dr. Angel Catalá was born in Rodeo (San Juan, Argentina). He studied chemistry at the Universidad Nacional de La Plata, Argentina, where received a Ph.D. degree in chemistry (Biological Branch) in 1965. From 1964 to 1974, he worked as Assistant in Biochemistry at the School of Medicine, Universidad Nacional de La Plata, Argentina. From 1974 to 1976, he was a Fellow of the National Institutes of Health (NIH) at the University of Connecticut, Health Center, USA. From 1985 to 2004, he served as a Full Professor of Biochemistry at the Universidad Nacional de La Plata, Argentina. He is Member of the National Research Council

(CONICET), Argentina, and Argentine Society for Biochemistry and Molecular Biology (SAIB). His laboratory has been interested for many years in the lipid peroxidation of biological membranes from various tissues and different species. Professor Catalá has directed twelve doctoral theses, published over 100 papers in peer reviewed journals, several chapters in books and, twelve edited books. Angel Catalá received awards at the 40th International Conference Biochemistry of Lipids 1999: Dijon (France). Winner of the Bimbo Pan-American Nutrition, Food Science and Technology Award 2006 and 2012, South America, Human Nutrition, Professional Category. 2006 award in pharmacology, Bernardo Houssay, in recognition of his meritorious works of research. Angel Catalá belongs to the Editorial Board of Journal of lipids, International Review of Biophysical Chemistry, Frontiers in Membrane Physiology and Biophysics, World Journal of Experimental Medicine and Biochemistry Research International, W orld Journal of Biological Chemistry, Oxidative Medicine and Cellular Longevity, Diabetes and the Pancreas, International Journal of Chronic Diseases & Therapy, International Journal of Nutrition, Co-Editor of The Open Biology Journal.

INDEX

D

E

F

G

H

I

K

L

N

O

P

Q

R

S

W